CAMBRIDGE LIBRARY COLLECTION

Books of enduring scholarly value

Mathematics

From its pre-historic roots in simple counting to the algorithms powering modern desktop computers, from the genius of Archimedes to the genius of Einstein, advances in mathematical understanding and numerical techniques have been directly responsible for creating the modern world as we know it. This series will provide a library of the most influential publications and writers on mathematics in its broadest sense. As such, it will show not only the deep roots from which modern science and technology have grown, but also the astonishing breadth of application of mathematical techniques in the humanities and social sciences, and in everyday life.

Memoirs of the Analytical Society

By the end of the eighteenth century, British mathematics had been stuck in a rut for a hundred years. Calculus was still taught in the style of Newton, with no recognition of the great advances made in continental Europe. The examination system at Cambridge even mandated the use of Newtonian notation. As discontented undergraduates, Charles Babbage (1791–1871) and John Herschel (1792–1871) formed the Analytical Society in 1811. The group, including William Whewell and George Peacock, sought to promote the new continental mathematics. Babbage's preface to the present work, first published in 1813, may be considered the movement's manifesto. He provided the first paper here, and Herschel the two others. Although the group was relatively short-lived, its ideas took root as its erstwhile members rose to prominence. As the society's sole publication, this remains a significant text in the history of British mathematics.

Memoirs of the Analytical Society

CHARLES BABBAGE
JOHN HERSCHEL

CAMBRIDGE
UNIVERSITY PRESS

University Printing House, Cambridge, CB2 8BS, United Kingdom

Published in the United States of America by Cambridge University Press, New York

Cambridge University Press is part of the University of Cambridge.
It furthers the University's mission by disseminating knowledge in the pursuit of
education, learning and research at the highest international levels of excellence.

www.cambridge.org
Information on this title: www.cambridge.org/9781108062404

© in this compilation Cambridge University Press 2013

This edition first published 1813
This digitally printed version 2013

ISBN 978-1-108-06240-4 Paperback

MEMOIRS

OF THE

ANALYTICAL SOCIETY

1813.

CAMBRIDGE:

Printed by J. Smith, Printer to the University;

AND SOLD BY

DEIGHTON & SONS, CAMBRIDGE;   LONGMAN & CO. PATERNOSTER-ROW, LONDON;

PARKER, OXFORD; CONSTABLE & CO. EDINBURGH; AND

GILBERT & HODGES, DUBLIN.

1813.

CONTENTS.

———◆———

PREFACE.

To examine the varied relations of necessary truth, and to trace through its successive developements, the simple principle to its ultimate result, is the peculiar province of Mathematical Analysis. Aided by that refined system, which the ingenuity of modern calculators has elicited, and to which the term Analytics is now almost exclusively appropriated, it pursues trains of reasoning, which, from their length and intricacy, would resist for ever the unassisted efforts of human sagacity. To what cause are we to attribute this surprising advantage? One, undoubtedly the most obvious, consists in the nature of the ideas themselves, whose relations form the object of investigation—and the accuracy with which they are defined. This is equally indeed the property of every branch of Mathematical enquiry. Three causes however chiefly appear to have given so vast a superiority to Analysis, as an instrument of reason. Of these, the accurate simplicity of its language claims the first place. An arbitrary symbol can neither convey, nor excite any idea foreign to its original definition. This immutability, no less than the symmetry of its notation, (which should ever be guarded with a jealousy commensurate to its vital importance,) facilitates the translation of an expression into common language at any stage of an operation,—disburdens the memory of all the load of the previous steps,—and at the same time, affords it a considerable assistance in retaining the results. Another, and perhaps not less considerable cause, is to be found in the conciseness of that notation. Every train of reasoning implies an exercise of the judgement, which, being an operation of the mind, deciding on the agreement or disagreement of ideas successively presented to it, it is reasonable to presume will be more correct, in proportion as the ideas compared follow each other more closely; provided the succession be not so rapid as to cause confusion. Were an Analytical operation of any complexity converted into common language, in all its detail, the mind, after acquiring a clear conception of one part of the related ideas, must suspend its decision until it could obtain an equally perspicuous one of the remainder of the proposition; and in so long an interval as this must occupy, the impression of the former ideas would necessarily have faded in some degree from the memory, unless fixed by an expense of time and attention, sufficient to deter any one from the employment of such means of discovery. It is the spirit of this symbolic language, by that mechanical tact, (so much in unison with all our faculties,) which carries the eye at one glance through the most intricate modifications of quantity, to condense pages into lines, and volumes into pages; shortening the road to discovery, and preserving the mind unfatigued by continued efforts of attention to the minor parts, that it may exert its whole vigor on those which are more important.

The last cause we have occasion to notice is, that Analysis, by separating the difficulties of a question, overcomes those which appear almost insuperable when combined, or at least, reducing each to its least terms, leaves them as the acknowledged landmarks of its progress,—open to approach on all sides, should ulterior discovery furnish any rational hope of their removal. Meanwhile that progress continues unimpeded. Simple relations are found to exist between the most refractory functions, and even when the difficulties themselves prove invincible, their nature at least becomes thoroughly understood, and means of evading them almost universally pointed out.

That the preceding observations are not founded on bare speculation, the whole history of Analytical Science will abundantly evince. It is our intention, in the following pages of this Preface, to give a general outline of that history up to the present time. From the space allotted to it, it is evident that little else than the most prominent points in so wide a field can be selected for observation. Faint, however, as it is, the subject cannot but communicate to it some portion of its interest; as well as the reflection, that, (with the exception of one branch of it *) the history of the more modern discoveries has hitherto unfortunately found little place in our language †.

Symbolic reasoning appears to have been ushered into the world under unfavourable auspices, and to have been regarded in its infancy with an eye of extreme jealousy. And, indeed, if we consider the rudeness of its first attempts, the poverty of its first resources, and the lavish want of œconomy in their employment, we shall find little reason to wonder, that for a long period, the new methods were looked upon as inelegant, although serviceable auxiliaries of the ancient processes, to be regularly discarded after serving their turn. To employ as many symbols of operation and as few of quantity as possible, is a precept which is now found invariably to

---

* The calculus of variations, the history of whose rise and progress has been ably combined with the exposition of its theory, in a late work, " On Isoperimetrical Problems."

† The admirable review of the Mecanique Celeste (Ed. Rev. Nº 22.) will still be fresh in the minds of our readers. But it should be recollected, that the Author of that Essay confines his attention entirely to the subject of Analytical dynamics; referring to the discoveries in the integral calculus merely as connected with that subject, and that too very cursorily. *Our business is exclusively with the pure Analytics.*

ensure elegance and brevity. The very reverse of this principle forms the character of symbolic analysis, up to within fifty years of the present date.

The first and most natural object of research in the Algebraic calculus, was the resolution of equations, involving simply the powers of the unknown quantity. As far as the fourth degree no difficulty occurred, but beyond this, not a step has yet been made. Almost every Analyst of eminence has applied his ingenuity to the accomplishment of the *general* problem, but without success; and after more than two centuries, during which every other branch of Analytics has been advancing with unrivalled rapidity, no progress whatever has been made in this. This, it must be allowed, presents little prospect of success to future researches on the subject; yet ought not the difficulty to be considered insurmountable, until opinion has been confirmed by demonstration. Delambre notices a Memoir presented to the National Institute by M. Ruffini, in which he proposes a proof of the impossibility of the resolution of equations above the fourth degree. If this demonstration be correct, it will render an important service to Algebraists, by diverting them from a pursuit which must necessarily be unsuccessful. The work, however, if yet published, has not arrived in this country. Recent French publications are not easily procured, nor is it surprising that to obtain those of the German Analysts is almost impossible, when Delambre regrets their scarcity even in France.

Although to express in finite algebraic terms, the root of any proposed equation be impracticable, yet the inverse function of any expression, such as

$$a + b\,x + c\,x^2 + d\,x^3 + \&c.$$

may readily be exhibited in an infinite series. When the difficulty of solving equations above the fourth degree was perceived, it was natural to seek rapid and convenient approximations, and accordingly, three of our countrymen, Newton, Ralphson, and Halley, produced nearly at the same time, modes of approximation which have since received various improvements. All such researches, when symbolically conducted, and without regard to the numerical value of the symbols, lead at the bottom to series of greater or less complexity. It seems to have been in following up this idea, that Lagrange was first conducted to that very general resolution of all equations in the series which bears his name; a series which has been productive of discovery wherever it has been applied, and whose fecundity appears yet far from being exhausted. It is thus that the most distant parts of Analysis hang together, nor is it possible to assign the point, however remote or unexpected, in which any proposed career of research may not ultimately terminate. This series made its first appearance in the Mem. de l'Acad. Berlin, 1767-8 *. together with

---

* The demonstration there given is defective in rigour. A better was given in Note XI. of the " Traitè de la resolution des equations numeriques." Lagrange. But the most elegant is that of Laplace,

the most systematic method of approximating to the roots of numerical equations which has yet been given. The method has this considerable advantage over all others, that, in all cases where the root is an integer, the formula of approximation will give it exactly, and in many where it is a surd, the continued fraction employed will point out the rational number of which it is the root. Though the complete solutions of equations is nearly hopeless, it might perhaps be of some advantage, and certainly of little difficulty, supposing the roots of equations known, to investigate what change would take place if one or more of the coefficients were augmented or diminished by any small quantity.

To trace the history of the differential calculus through the cloud of dispute and national acrimony, which has been thrown over its origin, would answer little purpose. It is a lamentable consideration, that that discovery which has most of any done honour to the genius of man, should nevertheless bring with it a train of reflections so little to the credit of his heart.

Discovered by Fermat, concinnated and rendered analytical by Newton, and enriched by Leibnitz with a powerful and comprehensive notation, it was presently seen that the new calculus might aspire to the loftiest ends. But, as if the soil of this country were unfavourable to its cultivation, it soon drooped and almost faded into neglect, and we have now to re-import the exotic, with nearly a century of foreign improvement, and to render it once more indigenous among us.

The most prominent feature of this calculus, is the theory of the developement of functions. The theorem which has immortalized the name of Brook Taylor, forms its foundation. Elicited by its Author from a formula which at first sight seemed independent of it, by a method not remarkable for its rigour, it seems to have been long considered in the light of a very general formula of interpolation. Lagrange and Arbogast have, as it were, invented it anew, and established it as the true basis of the differential calculus. The theory of Lagrange is to be found in a Memoir among those of the Acad. de Berlin. 1772, which contains the independent demonstration of Taylor's theorem—in the " Theorie des fonctions Analytiques," wherein he exhibits its application to the various branches of the differential calculus, independently on any consideration of limits, infinitesimals or velocities — and lastly, in the Journal de l' Ecole Polytechnique. Cah. XII. (1802.) and in the " Lecons sur le calcul des Fonctions." The ideas of Arbogast are contained in a Manuscript

---

to be found in the Mem. de l' Acad. des Sciences, Paris, 1777 ; in Lacroix's Calc. Diff. et Int. 2d edit. Art. 107 ; in the Mecanique Celeste, tom. I. page 172. Adopted (in principle at least) by Lagrange in the " Theorie des Fonct. Analyt. Art. 97, et suiv. And in our own language, in Mr. Woodhouse's Trigonometry, 2nd edition. Arbogast has also demonstrated this theorem. See his Calcul des Derivations, Art. 282, et suiv.

presented to the Academy of Sciences in the year 1789, and of which the outline is given in the Preface to his celebrated work on Derivations. Such is the brief account of the greatest revolution which has yet taken place in Analytical Science.

The operations of the differential calculus once well understood, and rigorously demonstrated, may be employed in improving the theory which gives rise to them. The work of Arbogast just alluded to, has shewn to how vast an extent this application may be carried, and how great is the assistance thus rendered. The peculiar grace of Laplace's Analysis has no where been more beautifully exhibited, than in his improvement and extension of Lagrange's theorem already mentioned. Nor should the labours of Paoli in this field pass unnoticed. By the aid of a very remarkable series derived from reverting that of Taylor, he has been able to assign the developement of any function of a quantity given by any equation whatever, in terms of a function any how composed of the remaining symbols which enter into that equation.

The developement of functions has lately been made, under the name of "Calcul des fonctions generatrices," the foundation of a most elegant theory of finite differences, of which more hereafter.

Soon after the discovery of the integral calculus, on the discussion of some problems, between Leibnitz and the Bernouillis, respecting the variation of the parameters of curves; there occurred certain equations, which, though they satisfied the conditions, were yet not contained in the complete integral of the equation whence they were derived. It is somewhat remarkable, considering the manner in which they first appeared, that their geometrical signification should have remained so long undiscovered. Brook Taylor, according to Lagrange, was the first who arrived at a particular solution by differentiating. Clairaut, in a Memoir presented to the Academy of Sciences at Paris, first remarked, that the equations so found, satisfy the geometrical conditions proposed. Euler styled them Analytical paradoxes, and shewed how in some cases they might be derived from the differential equations. But their theory remained unknown, till the year 1772 [*]; when Laplace explained it in a Memoir of the Academy of Sciences, and pointed out the methods of discovering all the particular solutions of which an equation admits. This subject was still farther pursued in the Berlin Memoirs, by Lagrange, who there developes, with great perspicuity the whole theory both Analytical and

---

[*] J. Trembley, in the 5th Vol. of the *Mem. de Turin*, has given a Paper on the derivation of the complete integral, having given a number of particular solutions. His method consists in multiplying the equation by these solutions, each raised to an indeterminate power.

Geometrical *. But the most complete exposition of the subject, which has yet appeared, is to be found in a paper read before the National Institute in the year 1806, by Poisson, in which the theory is extended to partial differential equations, and also to those of finite differences. He observes of certain partial differential equations, that they admit of particular solutions equally general with the complete integral. The Analytical theory, in its present state, is most elegant: still it requires some farther developements when applied to equations of partial differentials, and to those of differences, and might perhaps with advantage be applied to equations of mixed differences. Its Geometrical signification has been beautifully illustrated by Lagrange; but the meaning of particular solutions, when they occur in dynamical problems, which is a question of considerable importance, remains yet undecided. Poisson has shewn a case, in which the particular solution and the complete integral are both required, and has produced others, in which only one is necessary.

As the integration of expressions containing one variable is a matter of considerable importance, and the number of those which are capable of integration, is small, when compared with those which do not admit of it; some attention has been bestowed on the classification of those which are similar, and on the reduction of those which are absolutely different to the least number possible. When this is accomplished, all that remains for the perfection of this branch of Analysis, is to calculate tables which shall afford a value of the integral for any value of the variable. In general, all expressions which do not admit of complete integration, are denominated transcendants. Those which most frequently occur, are logarithmic and circular functions. Tables of these had been long calculated for trigonometrical purposes, and on the discovery of the integral calculus received a vast addition to their utility. It was next proposed to admit as known transcendants, all integrals which could be reduced to the rectification of the conic sections. But, besides the preposterous idea of limiting an Analytical expression by the properties of a curve, no tables had been constructed for them, and of course the determination of their arcs could only be performed by the actual calculation of the integral under consideration : nor, indeed, would it have been possible to form useful tables of any moderate length, without first discussing the properties of the transcendants themselves in the fullest manner.

The theory of the transcendant $\int \dfrac{P\, dx}{R}$, where P is a rational and integral function of x, and R a quadratic radical of the form $\sqrt{(\alpha + \beta x + \gamma x^2 + \delta x^3 + \epsilon x^4)}$ has at length by the successive labours of Fagnani, Euler, Landen, and Legendre,

---

* Fontaine first considered a differential equation as the result of the elimination of a constant between an equation and its differential; thus laying the foundation of the theory of equations, both differential, and of differences, and also of their particular solutions.

been brought to great perfection. All the transcendents comprised in this extensive formula, are reducible to three species. Those comprised in the first, are susceptible of multiplication or division, in the same way as the arcs of circles, by algebraic operations only. The transcendants of the second species, are susceptible of a similar multiplication or division, not simply, but when increased or diminished by an algebraic quantity. This algebraic quantity passes in the third species, into a transcendant of the logarithmic or circular kind. Landen has shewn, that an integral of the first species here enumerated may be reduced to two of the second: so that the number of distinct transcendants comprised in the formula $\int \frac{\mathrm{P}\, dx}{\mathrm{R}}$ is no more than two. A vast variety of integral formulæ have, by dint of indefatigable research on all hands, been reduced to the evaluation of these functions; but to dwell longer on them, would lead us beyond our limits.

The only other species of transcendants of any considerable extent which have received much discussion, are those contained in the formulæ $\int \varepsilon^{-x^{n}} .\, dx$, and $\int \frac{^{n}\varepsilon^{x} .\, dx^{n}}{1 + \varepsilon^{x}}$. Kramp, at the end of his Analyse des Refractions, has given a table of the values of the first of these, (in the case of $n = 2$), the integral being taken between the limits $0\cdot00, \ldots.3\cdot00, \infty$. The definite integrals dependent on the general form, we shall speak of hereafter. The second formula is (ultimately) that of the logarithmic transcendants, on the various orders of which Mr. Spence, in the year 1809 *, published an Essay, which displays considerable ingenuity, and a depth of reading rarely to be met with among the Mathematical writers of this country. A general property there given of the transcendant $^{n}\mathrm{L}\,(x)$, leads to the summation of some very extraordinary series, which are now in our possession, and which we cannot forbear mentioning.

Their general (or i^{th}) terms are comprised in the formulæ

$$\frac{f\{x.\varepsilon^{i\theta \sqrt{(-1)}}\} + f\{x.\varepsilon^{-i\theta\sqrt{(-1)}}\}}{i^{2n}}, \quad \text{and} \quad \frac{f\{x.\varepsilon^{i\theta \sqrt{(-1)}}\} - f\{x.\varepsilon^{-i\theta\sqrt{(-1)}}\}}{\sqrt{(-1)}.\, i^{2n+1}}$$

where f is the characteristic of any function whatever †, developeable in integer powers, either positive or negative, or both.

---

* Le Gendre published his Exercises de Calcul Integral in 1811. After having given Landen's and Euler's values of particular cases of the function $^{n}\mathrm{L}\,(1+x)$ he adds, " Jusqu'a present, on n'est pas allé plus loin dans la theorie de ces sortes des transcendantes," page 249. It is probable therefore, that, owing to our interrupted intercourse with the continent, he had not seen Mr. Spence's work.

† One singular result of these researches is, the evaluation in terms of the transcendants ϵ, and π, of the function

$$(\tan \theta)^{(\frac{1}{7})^{2n+1}} .\, (\cot 3\,\theta)^{(\frac{1}{3})^{2n+1}} .\, (\tan 5\,\theta)^{(\frac{1}{5})^{2n+1}} .\, \&c. \text{ ad infinitum.}$$

Mr. Spence has also given tables (of some extent,) of the successive values of his functions, as we have before remarked of Kramp. Too much praise cannot be bestowed on such examples, which however there is little hope of seeing followed. The ingenious Analyst who has investigated the properties of some curious function, can feel little complaisance in calculating a table of its numerical values; nor is it for the interest of science, that he should *himself* be thus employed, though perfectly familiar with the method of operating on symbols; he may not perform extensive arithmetical operations with equal facility and accuracy; and even should this not be the case, his labours will at all events meet with little remuneration.

It sometimes happens, that the arbitrary constant does not continue the same throughout the whole extent of an integral. An instance of this, in the series

$$C - \frac{\theta}{2} = \frac{\sin \theta}{1} + \frac{\sin 2\theta}{2} + \&c.$$

was remarked by Daniel Bernouilli, in the Act. Petrop. Landen also notices, that this equation is false, when $\theta = 0$, but without explanation. Other instances occur in the " Essay on logarithmic transcendants," pp. 52-3; in Lacroix's Traite de Calcul &c. 4to. tom. III. p. 141, as well as some reflections on this difficult subject in page 483 of the same volume. The cause of these anomalies has not been satisfactorily explained. If we may hazard a conjecture, it must be looked for in the evanescent or infinite values of some of the differential coefficients of the function integrated, causing that function for an instant to change its form. Or, it may have some connection with exponentials, since all the instances which have hitherto been adduced depend on that species of function.

The discovery of partial differentials, has been generally attributed to D' Alembert. He certainly was the first who applied them to mechanical problems, and perceived their vast utility in all the more difficult applications of Analysis to physics. But, if he is to be considered as the inventor, who first solved an equation of the kind, and who, when their importance was acknowledged, contributed more than any other to the improvement and progress of this calculus; the glory of their discovery will undoubtedly belong to Euler. In the 7th vol. of the Acta Acad. Petropolitanæ, is a Memoir of his, entitled, " Methodus inveniendi æquationes pro infinitis curvis ejusdem generis." (A. D. 1735.) In this paper, and more particularly in a supplement, are given the solutions of a number of partial differential equations, of which the most general is

$$q = X z + p R$$

X being a function of x, and R a function of x and y.

In the latter part of the " additamentum ad dissertationem," he proceeds to integrate some equations of the second order. The " Reflexions sur la cause generale des Vents," which contains d' Alembert's first application of partial differential equations, was not published till the year 1747, and it gained the prize of the Academy of Berlin in 1746. It would lead us too far to trace, successively, the various improvements which the new theory underwent in the hands of Euler, Lagrange, Laplace, Monge, Parseval, and a multitude of great men, whom the vast importance of the subject incited to its prosecution. Notwithstanding every exertion, the theory however continues to present a multitude of difficulties. The analogy which was supposed to exist between the arbitrary functions, which enter into the integrals of partial differential equations, and the arbitrary constants in equations of total differentials, is found not to hold beyond the first degree; after which, even the number of these arbitrary functions is unknown. Instances have been adduced, where, besides the arbitrary functions, arbitrary constants also must be introduced to complete the integral*. The application of definite integrals to the integration of these equations, presents a wide field of research, as well as the promise of great discoveries. A very curious Memoir of Laplace is to be found in the Mem. Acad. des Sciences 1779, where this subject, among many others, is discussed with considerable success †.

In applying the test of integrability to differential equations, some were found, which could not be made to satisfy the equations of condition. These were for a long time deemed absurd, until about the year 1784, when Monge perceived their connection with the theory of curve surfaces, and demonstrated that these equations admit of solutions, corresponding to the curves of double curvature, formed by the successive intersections of a curve surface, whose parameter varies; and discovered methods of transforming any equation of this kind into an equation of partial differentials, and also of solving the converse problem. From this, he proposed obtaining solutions of equations of partial differentials, which are not integrable by other methods. But the example he gives, shows, that to expect success in such an enquiry, we must be familiar with space, considered as of three dimensions, and also with a numerous collection of curves and curve surfaces situated in it; such considerations would but add intricacy to a subject already difficult: hints for the advancement of Analysis may be derived from foreign sources, but must always be improved and cultivated by its own powers. La Croix has given an excellent Analytical theory of this kind of equations.

To the practice which prevailed in the infancy of Analytics, of proposing to

---

* Monge; Savans Etrangers, vol. VII. p. 322.

† The great desideratum in the integration of an equation by definite integrals is, that whenever it is susceptible of actual resolution, these integrals should give it—a condition which does not always hold good. See Laplace. Mémoire sur divers points d' Analyse. Journ. de l' Ecole Polytechnique, N°. 15.

Geometers the solution of difficult problems, we owe many of its improvements. The method of variations, among others, is much indebted to this source. Obliged however to consult brevity, we must refer to the late treatise on " Isoperimetrical Problems," for a full account of its history: a work which being undoubtedly in the hands of the majority of our readers, must render superfluous all we could say on that subject. The chief difficulty now consists in distinguishing the maxima from the minima, which depends in general on the solution of difficult differential equations. The inverse of the method of variations does not appear to have received any attention; it would consist in solving such problems as the following: " Given a curve, to find what properties of maxima and minima it possesses."

The method of variations was applied by its inventor to differential equations, and also to those of finite differences. Cases may occur, in which it would be necessary to apply it to equations of mixed differences; these relate to a number of difficult problems, such, for instance, as this: " What must be the nature of a curve, such, that drawing to any point, an ordinate and also a normal, and at the foot of this normal another ordinate; the curvilinear area intercepted between the first and last ordinate, may be a maximum or a minimum."

There are but few instances in the history of Science, in which the path of the inventor has been the shortest and most direct. Thus it occurs, that the method of finite differences, which would most naturally have preceded that of the differential calculus, was not discovered until many years after. Its inventor, Brook Taylor, published it in a work, entitled, " Methodus Incrementorum directa et inversa," a book noted for its obscurity. Montmort, Stirling, and Emerson, made several improvements, which may be found in the Philosophical Transactions, and also in their respective works. Moivrè first investigated the nature of recurring series, on which, Laplace remarks, " Sa theorie est une des choses les plus curieuses, et les plus utiles que l'on ait trouvees sur les suites." It has been well observed, by the same author, that the first who summed a Geometrical or Arithmetical series, had really integrated an equation of finite differences. The same remark, as is well known, applies to any recurrent series. It was not, however, till Lagrange in the Melanges de Turin, vol. I. applied Alembert's method of indeterminate coefficients, to an equation of differences of the first degree, that this truth was perceived. In the fourth volume of the same work, Laplace published a Memoir, in which the two celebrated theorems of Lagrange, respecting equations of common differentials, are extended to those of differences, and to those of partial differentials of a similar description, with constant coefficients. Returning to the subject, in a Memoir communicated to the Academy of Paris, he integrates a very extensive class of equations of partial differences, involving any number of variable indices,—and also a singular species of equations, frequent in the theory of chances, called by him, *equations rentrantes*.

Of equations beyond the first degree, very few have been solved, if we consider their amazing variety and importance. Monge has given a short paper on the subject, in the Memoirs of the Academy of Sciences, for 1783. Laplace also in the 15th number of the Journal de l' Ecole Polytechnique, has by a most happy combination of the equations of differences, with the discovery of Euler, respecting elliptic transcendants, integrated a few very difficult ones. The nature of the integrals obtained by Charles, requires a fuller investigation. They might perhaps receive considerable extension. When the variables are mixed in the indices, thus, u_{x+y}, u_{x-y+1}, u_{xy}, &c.; the subject seems to have passed altogether unnoticed. Many equations containing such expressions, are impossible or contradictory.

Euler first remarked, that the constant introduced by integrating an equation between u_x, u_{x+1}, &c. may be an arbitrary function of $\cos 2\pi x$, a remark which afterwards in the hands of Laplace, (Savans Etrangers, 1773,) became the foundation of a very general theory of determining functions from given conditions. To notice all the applications of the theory of finite differences, or all the profound researches which have enriched it, would occupy volumes. We cannot, however, pass over the theorems relating to the analogy of differences to powers, given first by Lagrange without demonstration, in Mem. de Berlin. 1772. A demonstration by Laplace, appeared in the Mem. des Savans Etrangers for the following year, and another in the Mem. de l' Acad. for 1777. In 1779, appeared that noted Memoir, in which the same author exhibited the principles of his powerful and elegant " Calcul des fonctions generatrices." In this *, he gives a far more systematic proof of the theorems, and extends them to any number of variables. Since that period, they have been demonstrated by Arbogast, in the 6th article of his, " Calcul des derivations," where, by a peculiarly elegant mode of separating the symbols of operation from those of quantity, and operating *upon them* as upon analytical symbols; he derives not only these, but many other much more general theorems with unparalleled conciseness. Brinkley has given a demonstration of the theorem

$$\Delta^n u_x = \left\{ \varepsilon^{\frac{\Delta x}{d x} \cdot d \cdot} - 1 \right\}^n \times u_x$$

of considerable elegance, and a simplicity truly elementary †.

---

* This Memoir forms the greater part of the first Chapter of the " Theorie Analytique des Probabilites." Its first principles, and the demonstration of the theorems, on the analogy of differences with powers, are given briefly in the 15th N°. of the " Journal de l' Ecole Polytechnique." A slight sketch of the method alluded to, is also to be found in the 9th book of the Mécan. Cél. tom. IV. p. 204.

† Philos. Transactions, 1807. Part I. He has extended his researches to the actual expansion of the series themselves, to which these theorems lead; such, for instance, as

$$\Sigma u_x = \frac{1}{\Delta x} \cdot \int u_x \, dx - \frac{u_x}{2} + \Delta x \cdot {}^1 A_x + (\Delta x)^2 \cdot {}^2 A_x + \&c.$$

But in point of clearness and elegance, by no means with equal success: partly owing to an unfortunate notation, and partly to the perpetual employment of the theory of combinations. In his results, he has for the most part been anticipated by Laplace, and others.

We are now naturally led to say a few words on the " Calcul des fonctions gene-ratrices." Its object may be best stated, in the words of its inventor * : " C'est de ramener au simple developpement des fonctions, toutes les operations relatives aux differences et specialement l' integration des equations aux differences ordinaires et partielles," and from the extreme facility with which all the known theorems flow from it, and its fecundity in affording new ones, it is, perhaps, in the present state of science, the best adapted of any, for explaining the general theory of differences, and the developement and transformation of series. At the same time, it must be con-fessed, that owing to its extreme generality, and the consequent complexity of many of its operations, particularly in what regards the transformation of series, it is an instrument to be placed only in the hands of an experienced Analyst. If we except the calculus of variations, it is the only method, perhaps, of any considerable impor-tance, which has received its first and last touches from the same hand; and which first appeared in a state of perfection, very little short of what it at present possesses. The latest work which treats of this subject, is the " Theorie Analytique des Proba-bilite's;" the first part of which is dedicated to a very full exposition of the method.

After the solution of differential equations, and those of finite differences, it was natural to consider those, in which the difference and differential of a quantity both occurred. These have been called equations of mixed differences. They were not attempted until about the year 1779, when Condorcet and Laplace obtained the integrals of some few particular cases. But problems which required their application had been proposed and resolved by several Geometers in the " Acta eruditorum," and in the " Acta Acad. Petrop." long before this time; their solutions were obtained by certain insulated artifices, dependent on the peculiar nature of the problems.

On this subject, a wide field is extended for investigation, and one which abounds with difficulties. The little that has yet been discovered, is chiefly con-tained in two papers, one by Biot, in the *Memoires de l' Institut*, and the other by Poisson, in the *Journal de l' Ecole Polytechnique*; the former treats chiefly of that kind of equations, called *equations successives*; the latter integrates a few particular equations, by employing a substitution used by Laplace, in integrating some equations of partial differentials.

There is no branch of mathematical science which has not received improvements, from the profound and original genius of Euler. Several are indebted to him for their existence; of this latter class is the knowledge of the nature, and use of definite integrals, a subject, to which the greatest Geometers of the present age look as the most probable source of future discoveries and improvements.

---

* Journal de l' Ecole Polyt. Mém. sur divers points d' Analyse.

Legendre, in a work lately published, entitled " Exercises de Calcul Integral," has collected all that has been discovered on this subject, and demonstrated the results with peculiar elegance; the greater part is extracted from the various works of Euler; and also a considerable portion from some Memoirs of Laplace. Its application to the solution of differential equations, from which so much is expected, does not enter into the plan of his work; this however has been well treated, by Lacroix, in the third volume of his *Traite de Calcul, &c.*

But the most elegant part of the theory of definite integrals is, their application to such problems in finite differences, as involve functions of very high numbers. In many cases (particularly in the theory of chances,) it has been well remarked, that the mere impracticability of the arithmetical operations requisite to obtain a result, (however simple its analytical expression,) must for ever preclude our advancement; were it not for some mode of approximation, which, grasping the prominent terms of an expression, in a formula easily reduced to numbers, should throw the minor ones into the back-ground, to be valued by a series converging the more rapidly the higher the numbers employed become. A more appropriate instance cannot be adduced, than the equation

$$1.2.3.....s = s^{s+\frac{1}{2}}.\varepsilon^{-s}.\sqrt{(2\pi)}\left\{1 + \frac{1}{12s} + \frac{1}{288s^2} + \&c.\right\}$$

in page 129 of the " Theorie Anal. des Probabilites," or the method in which the author, in page 259 of the same work, computing the probability of a primitive cause influencing the inclinations of the cometary orbits, employs a definite integral to effect with conciseness, a calculation surpassing, without that assistance, the utmost limits of human patience and industry.

In this part of his career, Laplace stands unrivalled. Stirling indeed, and Moivre had seen, and in some cases obviated the difficulties, arising from the immensity of the numbers under consideration. The discoveries of Euler, gave a connection and unity to their results. But it was not till the labours of Lagrange, Condorcet, and Laplace had brought the theory of finite differences to considerable perfection, that the definite integrals were applied by the latter, to the solution of these equations, and a clear and strong light thrown over this most obscure part of the Mathematics. The whole of this interesting theory, has been digested into one work, (" Theor. de Prob." above cited) which for comprehensive views, for depth of investigation, and the purity of its analysis, may justly be looked up to, as marking the highest point to which the science of abstract number has yet attained.

In analytical investigations, we frequently meet with a series of quantities connected together by multiplication, whose differences are constant. These have received various names from different Geometers; Vandermonde called them powers of the second order; Kramp adopted the appellation of Facultes numeriques; and Arbogast named them Factorials. Each of these writers treats them in a different manner, and employs a peculiar notation for them: that of Vandermonde is perhaps the best adapted to the subject, as it has a considerable resemblance to that of exponents, and possesses also an advantage, which is by no means inconsiderable, that of being capable of a ready extension to powers of all orders. Lacroix has adopted it, and by its help demonstrated many properties of powers of the second order. Those of the superior orders have not as yet been examined. Kramp has deduced from his " Theorie des facultes numeriques," some contradictions which require examination. One of his most useful theorems affords a method of transforming any power of the second order, into a series which converges *ad libitum*. It appears probable, that the theory of powers of different orders may afford a useful method of classing transcendents, as they can frequently be reduced to definite integrals, and by this means their value be obtained, when their index is fractional.

Interpolations were at first considered, as a branch of the method of finite differences, and as such they were usually treated of together. Wallis appears first to have applied this name, to the determination of the intermediate term of a series, whose law of formation is known. The extraction of roots is an interpolation of powers, and may be considered as an extension of the meaning of exponents, from whole numbers to fractions. Perhaps it might not be unworthy of consideration, whether the meaning of the indices of differentiation, could not be considerably extended. Euler seems to have had the first idea* of interpolating the series

$$dy, \quad d^2y, \quad d^3y, \quad \&c.$$

Laplace has extended his researches on this subject to considerable length, and has given the value of such expressions as the following, by converging series and definite integrals

$$\frac{d^n . x^m}{dx^n}, \qquad \Delta^n . x^m$$

so as to allow of evaluation for fractional values of n. But the indices themselves might be supposed to vary *continuously*, and such expressions as these

$$\left\{ \frac{d^n . \left(\frac{d^p y}{dx^p} \right)}{dp^n} \right\}, \qquad \left(\frac{d \Delta^n y_x}{dn} \right)$$

become the subject of Analytical investigation. Or the index of a function might vary, as in the following instance:

---

* Leibnitz, it is true, in a letter to J. Bernouilli, mentions fractional indices of differentiation. Euler, however, first determined the value of such an expression as $d^{\frac{1}{2}}y$, y being a certain function of x.

Let $\quad f(x) = \sqrt{\left(\dfrac{2x}{1+x}\right)}, \qquad f^2(x) = \sqrt{\left(\dfrac{2f(x)}{1+f(x)}\right)}, \quad$ and so on,

then shall we have,

$$\frac{d f^n(\sec v)}{d n} = -\frac{v \cdot \log 2}{2^n} \cdot \tan\left(\frac{v}{2^n}\right) \cdot \sec\left(\frac{v}{2^n}\right)$$

and again, (making use of Arbogast's notation,) if $y = a^x$

$$\mathrm{D}_{p_{n-1}}^{p_n} \cdot \mathrm{D}_{p_{n-2}}^{p_{n-1}} \cdot \ldots \cdot \mathrm{D}_{p_1}^{p_2} \cdot \mathrm{D}_x^{p_1} \cdot y = a^x \cdot (\log a)^{p_1} \cdot (\log^2 a)^{p_2} \cdot \ldots \cdot (\log^n a)^{p_n}$$

where $\log^2 a = \log \log a$, $\quad \log^3 a = \log \log \log a$, and so on.

It has been observed by Charles, that the equation

$$\Delta y_n = b \cdot \frac{d y_n}{d x}$$

may be transformed into an integral, in which the index of integration is variable. Its solution then is

$$y_{-n} = b^{-n} \varepsilon^{-\frac{x}{b}} \cdot \int^n \varepsilon^{\frac{x}{b}} \cdot y_0 \cdot d x^n$$

or, which comes to the same,

$$y_n = b^n \varepsilon^{-\frac{x}{b}} \cdot \mathrm{D}_x^n \left\{ y_0 \cdot \varepsilon^{\frac{x}{b}} \right\}$$

y_0 being any function of x. Laplace, in the " Theorie Anal. des Prob." gives other instances of the same kind *.

That such expressions are not merely analytical curiosities, but relate to the most difficult and important theories, is confirmed by the opinions of the most

---

* The integral of the equation of mixed partial differences

$$u_x = \mathrm{A} \cdot \left(\frac{d^\alpha u_{x-1}}{d a^\alpha}\right) + \mathrm{B} \cdot \left(\frac{d^\beta u_{x-2}}{d b^\beta}\right) + \mathrm{C} \cdot \left(\frac{d^\gamma u_{x-3}}{d c^\gamma}\right) + \&c.$$

may be easily shown to be

$$u_x = \mathrm{A}^{\frac{x}{1}} \cdot \left(\frac{d^{\frac{\alpha x}{1}} \cdot \psi_1(a)}{d a^{\frac{\alpha x}{1}}}\right) + \mathrm{B}^{\frac{x}{2}} \cdot \left(\frac{d^{\frac{\beta x}{2}} \psi_2(b)}{d b^{\frac{\beta x}{2}}}\right) + \mathrm{C}^{\frac{x}{3}} \cdot \left(\frac{d^{\frac{\gamma x}{3}} \psi_3(c)}{d c^{\frac{\gamma x}{3}}}\right) + \&c.$$

ψ_1, ψ_2, &c. being the characteristics of arbitrary functions : an expression which except α, β, γ, ... are respectively multiples of 1, 2, 3, ... must necessarily involve fractional indices of differentiation for some values of x.

eminent Analysts ; and though on mathematical subjects when proof can be produced, no weight must be allowed to authority; yet when the former is deficient, our judgement may surely be influenced by the latter.

The importance of adopting a clear and comprehensive notation did not, in the early period of analytical science, meet with sufficient attention ; nor were the advantages resulting from it, duly appreciated. In proportion as science advanced, and calculations became more complex, the evil corrected itself, and each improvement in one, produced a corresponding change in the other. Perhaps no single instance of the improvement or extension of notation, better illustrates this opinion, than the happy idea of defining the result of every operation, that can be performed on quantity, by the general term of function, and expressing this generalization by a characteristic letter. It had the effect of introducing into investigations, two qualities once deemed incompatible, generality and simplicity. It now points out a calculus * perhaps more general than any hitherto discovered, and which should be called the calculus of functions, a name that more naturally belongs to it, than to that which Lagrange has so classically treated in the work which bears this name, although this latter is a branch of it.

Its object would be in general, the determination of functions from given conditions of whatever nature, whether depending on the successive terms of their developements, or on a series of indices differing by unity; or lastly, on a species of equations depending on the successive *orders* of the same function, of which the first mention we believe is made in one of the papers which compose the present volume. The necessity of this calculus was perceived soon after the discovery of equations of partial differentials, when it became requisite to determine the arbitrary functions which enter into their integrals, so as to satisfy given conditions. Euler and Alembert determined a few particular cases. Lagrange, in a Memoir entitled, " Solution de differens Problemes de calcul integral," resolved the equation

$$\mathbf{T} = a \cdot \phi \{t + a \cdot (h+kt)\} + \beta \cdot \phi \{t + b (h+kt)\} + \&c.$$

Monge also has given several papers upon the subject, in the Memoirs of the Societies of Turin and Paris. The method of Laplace for reducing an equation of the first order, where the difference of the independent variable is any function of the variable itself, to one wherein that difference is constant, is well known. It had not, however, hitherto been shewn to be possible to reduce *every* equation of the former kind, to one of the latter. This object is, however, accomplished in the following pages. Still, it appears by no means natural, to resolve these equations, by

---

* Quamobrem non solum in hoc negotio, sed in plurimis aliis casibus, maximé utile foret, si functionum doctrina magis perficeretur et excoleretur.   Euler, Act. Acad. Petropol. tom. VII.

means of finite differences, and it were much to be wished, that some independent method could be discovered, by which they might be treated.

One of the most striking advantages of the theory *of functions*, is, that it seems equally adapted to the proof of the most elementary truths, and to that of the most complicated and abstruse theorems. The latter part of this assertion, no one will be inclined to deny. An example of the former may be found in Laplace's proof of the decomposition of forces in the "Mecanique Celeste" *.

There are still many problems in the theory of functions, which analysis seems

---

* This demonstration consists of two parts. In the first he proves, that the diagonal of a rectangle whose sides represent the separate forces, will on the same scale represent the *quantity* of the resultant. The second is devoted to shew, that it represents also its direction. This part seems to be generally considered, as deficient in clearness and simplicity. What we have here to remark, is, that it is *redundant*. In fact, by the combination of Laplace's three equations,

$$x = z \cdot \phi(\theta); \quad y = z \cdot \phi\left(\frac{\pi}{2} - \theta\right), \quad x^2 + y^2 = z^2$$

we obtain

$$\left\{\phi(\theta)\right\}^2 + \left\{\phi\left(\frac{\pi}{2} - \theta\right)\right\}^2 = 1$$

an equation which suffices for determining the nature of the function ϕ, and from which, by known processes, combined with the conditions of the question; it is easy to obtain $\phi(\theta) = \cos\theta$, and $x = z \cdot \cos\theta$, which is the equation to be deduced.

Another very remarkable instance of the use of the theory of functions, in demonstrating elementary truths, may be found in the following demonstration of Euclid's 47th, which has generally been thought to admit of none, but a geometrical proof: Call a, b, c, the sides, A, B, C, the opposite angles of a right-angled triangle, $C = \frac{\pi}{2}$. It is easy to see, that the following equations hold good:

$$b = c \cdot \phi(A); \quad a = c \cdot \phi(B); \dots\dots\dots(1)$$

drop a perpendicular p, dividing c into two parts x, y, and we shall have, in the same manner,

$$x = b \cdot \phi(A); \quad y = a \cdot \phi(B)$$

and of course, $$x + y = c = b \cdot \phi(A) + a \cdot \phi(B)$$

from which, eliminating $\phi(A)$, and $\phi(B)$, by equations (1)

$$c^2 = a^2 + b^2, \quad Q E D.$$

Having proved from other principles, that $A + B + C = \pi$, we shall have the very same equation

$$1 = \left\{\phi(A)\right\} + \left\{\phi\left(\frac{\pi}{2} - A\right)\right\}^2$$

and thus we obtain $b = c \cdot \cos A$, $a = c \cdot \cos B$. It is only by this way of proceeding, or some analogous one, that we can ever hope to see the elementary principles of Trigonometry, brought under the dominion of Analysis. But this is not the place to proceed farther with the subject. It may suffice to have thrown out a hint, which may be followed up at some future opportunity.

e

to afford no means of attacking directly. Among which, may be enumerated the greater part of those which lead to an equation, containing a definite integral, where the unknown function enters under the integral sign. For instance, suppose it were required to find the form of a function $\phi(x)$, such that the integral

$$\int dx \, . \, \mathrm{F}\{a, x, \phi(f(x))\}$$

taken between the limits $x = 0$, $x = a$, should equal any assigned function of a. F and f being given characteristics.

The examination of the properties and relations of numbers, constitutes a distinct branch of mathematical enquiry, almost entirely of modern origin; so abstract, and apparently so far removed from the confines of utility, that it seems to have attracted little attention from the generality of those, who have dedicated themselves to the pursuits of science. To the few, however, who have thought it worth their while to explore its more profound recesses, it has proved a mine, fertile in the most brilliant produce. Euclid, in his 7th book, has given the elements of transcendental Arithmetic, (a name appropriated to it by Professor Gauss). In the work of Diophantus, notwithstanding the ingenuity of the author, we discover the infancy of science in the absence of that generalization so happily adopted by modern writers. It consists of a variety of insulated problems, relating to the solution of certain indeterminate equations, rather than to the properties of numbers. Indeed, the indeterminate analysis, and the theory of numbers, form two branches of enquiry, which, (however nearly connected), ought to be carefully distinguished, in any systematic arrangement of our knowledge. The former must be considered as a province of the pure Analytics. Its attainment is indeed necessary for the perfection of the latter, (which should rather be regarded in the light of an *application* of analysis,) but is by no means limited to this one object. Lacroix * has introduced it with a very elegant effect, in that part of the theory of curves, which relates to their construction by points.

The resolution of the indeterminate equation of the first degree, is said to be due to Bachet. Euler, in the Petersburg Commentaries, exhibited a method of obtaining any number of solutions of that of the second, provided one particular one be known; but it has been remarked, that his methods do not afford *all* the possible solutions. This, however, has been effected at length by Lagrange, in the Mem. Acad. de Berlin (1767 and 1768.) His method consists in reducing successively by a series of operations, the coefficients of the equation

$$a x^2 + b y^2 = z^2$$

(to which form, every equation of the second degree may be reduced,) till one of

---

* Traité de Calc. &c. 2d edit. vol. I. Note to page 417.

them becomes unity; in which case, the resolution is easy. Gauss also, in his " Disquisitiones Arithmeticæ," N° 216. has, by a method entirely different, shewn how to obtain all the solutions, of which an equation of the second degree admits. It is extremely remarkable, that the Hindu Algebraist, Bhaskara Acharya, who flourished about the year 1188, had also succeeded so far in his attempts on this difficult problem, as to derive any number of solutions from one, previously known, in the case of $a x^2 + b = y^{\circ}$.

The theorems given by Fermat without demonstration, form, without doubt, the most remarkable era in the theory of numbers. Too general, and too extraordinary in their nature to escape notice, they seem to have been the principal cause of the advances, which have since been made in this theory, by drawing the attention of Mathematicians to their demonstration. Euler, than whom none ever entered with greater ardency into this career, has proved some of the principal. Lagrange has supplied the demonstrations to others: still, however, many remain, of which no proof has been offered. It has been suggested, that Fermat was indebted to the method of induction, for the discovery of many of his theorems; an opinion rendered probable by the observation of Euler, that one of them relating to prime numbers is not true. This method is perhaps more applicable to researches in the theory of numbers, than to any other branch of abstract investigation; but it is of dangerous use, and should be supported by a large number of instances *.

---

* The substitution of 0, 1, 2, ... for x, in the expression $x^2 + x + 41$, gives a series of numbers, of which the 40 first terms are primes, as Euler has remarked: yet it is easy to shew, that no algebraic function of x can in all cases represent a prime. The reason of this singular circumstance, and of a variety of similar coincidences, has since been satisfactorily explained, and the property demonstrated *a priori*. Fermat, deceived by a similar induction, asserted that all the numbers contained in the formula $2^{\cdot n} + 1$, are primes which Euler has since (as above alluded to) shewn to fail, in the case of $n = 5$.

The following theorems are derived solely by induction:

1mo. An indefinite number of integer values of x may be found, which render $\dfrac{7^x - 1}{10^a}$ an integer, to which we may add, that the formula $\dfrac{7^{8 \cdot 10^x} - 1}{10^{a+2}}$ is always an integer, as are also the formulæ $\dfrac{3^{2^x} - 1}{2^x}$, $\dfrac{5^{2^x} - 1}{2^x}$ and, in general, $\dfrac{(2n-1)^{2^x} - 1}{2^x}$, which last may be easily shewn, *a priori*, as well as a variety of expressions of the same description.

2do. The expression $\dfrac{3^{10^x} - 1}{10^{x+1}}$ is an integer, and, it is somewhat remarkable, that this integer is always of the form $100 n + 22$.

3°. If A be such, that 5^n is congruous to A, (modul. 10^k) then will also $5^{n+2^{k-2} \cdot i}$ be congruous to the same A, to the same modulus, and consequently, $5^{n+2^{k-2} \cdot i} \equiv 5^n$. (modul. 10^k) i being any integer. In other words, the formula

$$\frac{5^{n+2^x \cdot i} - 5^n}{10^{x+2}}$$

is always an integer, x and i being integers.

Among the later discoveries in the theory of numbers, we have to enumerate two of the most surprising, which perhaps are to be found in the whole circle of Analytics. The first is a formula of Lagrange, obtained by induction, for determining the number of primes contained between given limits. It has been demonstrated by Legendre, in his " Essai sur la Theorie des Nombres," although not very rigorously, it must be confessed; such is the difficulty of the subject. Indeed, the author has given it only as an attempt. The other is that celebrated theorem of Gauss, given in his " Disquisitiones Arithmeticæ," on the resolution of the equation $x^n - 1 = 0$, n being any prime, viz. that this equation may be reduced to a equations of the degree a, β of the degree b, and so on, where $n - 1 = a^\alpha . b^\beta ....$ Of course, the resolution of the equation $x^n - 1 = 0$, where n is a prime of the form $2^m + 1$, requires only the application of quadratics. Thus the division of the circle into 17, 257, 65537 parts, may be accomplished by the description only of circles and straight lines.

To enter into any account of the advances made in the mixed Analytics, would far exceed our limits. There is one point, however, which we cannot forbear cursorily touching upon, on account of the great difficulty of reducing its conditions into symbolic language. We allude to the geometry of situation. Like the theory of numbers, at the first glance it seems barren and useless, but on a nearer examination is found abounding with interesting relations. Like that theory too, its cultivators have hitherto been few, but eminent, distinguished for that restless spirit of enquiry, which is ever upon the wing in search of new truths, and that invention which knows how to extract them, from the most unpromising hints. Leibnitz, appears to have found its first application, in considering the game of solitaire. A similar case (the problem of the knight's move at chess,) occupied the attention of Euler, and afterwards of Vandermonde, who adapted to it a notation analogous to that, by which the position of a point in space is determined, by three rectangular co-ordinates. In a more advanced state, it might, perhaps, embrace problems of a much higher order of difficulty, such as the following: " Given n points in space; to find the course to be pursued, so that setting off from any one, and passing at least once through all the rest, on returning to the original position, the least possible space shall have been described." Such is the brief account of a theory yet in its first infancy. On its basis some future LEIBNITZ may perhaps hereafter lay the foundation of a name great as that of its original inventor *.

---

* In the " Journal de l' Ecole Polytechnique, An. 10." is a Memoir on polygons and polyhedrons, by Poinsot, in which he shews, that there exist other regular polygons besides the equilateral triangle, the sum of whose angles is equal to two right angles, and also, that there are more than five regular polyhedrons. The whole Memoir relates to geometry of situation, and forms the introduction to some more considerable researches, which the author promises in a future paper.

The preceding pages have been devoted to a slight account of the history and present state of Analytical Science, that branch of human knowledge, of which Laplace has justly observed " C'est le guide le plus sur qui peut nous conduire dans la recherche de la verite." But some account will naturally be expected of the source itself, from which the present work emanates. Of this however, very little need be said, but, that it consists of a few individuals, perhaps too sanguine in their hopes of promoting their favourite science, and of adding at least some trifling aid to that spirit of enquiry, which seems lately to have been awakened in the minds of our country-men, and which will no longer suffer them to receive discoveries in science at second hand, or to be thrown behind in that career, whose first impulse they so eminently partook. The time perhaps is not far distant, when such an attempt will be regarded in an honourable light, whatever may be its success.

Meanwhile the view we have taken of the subject, appears by no means to lead to the mortifying conclusion, deduced by a foreign Geometer of considerable emi-nence ; " que la puissance de notre analyse est a-peupres epuisee." The golden age of mathematical literature is undoubtedly past. Another, " less fine in carat," may however yet succeed. The motive which could draw forth so severe a sentence on the success of future exertions we will forbear to enquire, but it must surely be looked for elsewhere, than in the real interest of science which can never be promoted by repressing the ardour of research, or extinguishing the hope of reward. The foundations of a vast edifice have been laid ; some of its apartments have been finished ; others yet remain incomplete : but the strength and solidity of the basis will justify the expectation of large additions to the superstructure.

Attentively to observe the operations of the mind in the discovery of new truths, and to retain at the same time those fleeting links, which furnish a momentary con-nection with distant ideas, the knowledge of whose existence we derive from reason rather than perception, are objects in whose pursuit nothing but the most patient assiduity can expect success. Powerful indeed, must be that mind, which can simultaneously carry on two processes, each of which requires the most concentrated attention. Yet these obstacles must be surmounted, before we can hope for the discovery of a philosophical theory of invention ; a science which Lord Bacon re-ported to be wholly deficient two centuries ago, and which has made since that time but slight advances. Probably, the era which shall produce this discovery is yet far distant. The capital of science, however, from its very nature, must continue to increase by gradual yet permanent additions ; at the same time that all such ad-ditions to the common stock yield an interest in the power they afford of mul-tiplying our combinations, and examining old difficulties in new points of view. It is this connection with fresher sources, which can restore fertility to subjects appa-

f

rently the most exhausted, and which cannot be too earnestly recommended to those who wish to enlarge the limits of analysis. The fire of improvement, however dormant, and seemingly extinct, may yet break forth at the contact of some external flame. The history of Mathematics affords too many instances of the most distant principles coming into play on the most unexpected occasions, to allow of our ever despairing of success in such enquiries.

One inconvenience however, results as a necessary consequence from the continued accumulation of indestructible knowledge. The beaten field of analysis, limited as it is when compared with the almost boundless extent which remains to be explored, is yet so considerable with respect to the powers of human reason, and (if we may be allowed to pursue the metaphor a little farther,) so intersected with the tracks of those who have traversed it in every direction, as to become bewildering and oppressive to the last degree. The labour of one life would be more than occupied in perusing those works on the subject which the labour of so many has been spent in composing. The multitude of different methods and artifices, which for the most part lead only to the same results, and whose power is limited by the same points of difficulty, is at length grown into a very serious evil. Our continental neighbours seem sensible of this, if we may judge from the number of works which have appeared within these few years, digesting various points into a systematic form. But there is still much to be done in this line. That man would render a most invaluable service to science, who would undertake the labour of reducing into a reasonable compass the whole essential part of analysis, with its applications, curtailing its superfluous luxuriance, rejecting its artificial difficulties, and giving connection and unity to its scattered members.

ERRATA.

Page 3. line 4. *for* exponent $\overline{p-1}$, *read* $-\overline{p-1}$

    ib. last line but one. *for* z_{x+1}, *read* u_{x+1}

    15. line 4. *for* 2 in the exponent, *read* -2

    24. last line. *for* $\log. \left\{ (1-x)^{\frac{1}{1-x}} \right\}^{1-x}$, *read* $\left\{ (1-x)^{\frac{1}{1-x}} \right\}^{1 \; x}$

CONTINUED PRODUCTS.

———

Assume the equation,

$$\psi x \cdot \phi x = \chi x \dots\dots\dots (a)$$

in which the functions denoted by ψ, ϕ, χ, are only subject to one condition, namely, that the product of the two first is equal to the last. In other respects they are perfectly arbitrary.

Take any other arbitrary function $f x$. It is evident that the generality of equation (a) is neither increased nor diminished by putting $\phi f x$ for χx.

Let this be done, then,

$$\psi x \cdot \phi x = \phi f x \dots\dots\dots (b)$$

In this equation put for x successively $f x$, $f f x$, $f f f x$, and $f^n x$; then multiply the resulting equations together,

$$\psi f x \cdot \phi f x = \phi f^2 x$$
$$\psi f^2 x \cdot \phi f^2 x = \phi f^3 x$$
$$\psi f^3 x \cdot \phi f^3 x = \phi f^4 x$$
$$\&\text{c.} \qquad \&\text{c.}$$
$$\psi f^n x \cdot \phi f^n x = \phi f^{n+1} x$$

therefore,

$$\frac{\phi f^{n+1} x}{\phi f x} = \psi f x \cdot \psi f^2 x \cdot \psi f^3 x \dots \psi f^n x \dots\dots (c)$$

In (b) we may without limiting the equation put $(\psi x)^{r-1}$ for ψx, it then becomes

$$(\psi x)^{r-1} \phi x = \phi f x \dots\dots\dots\dots (d)$$

A

Dividing by $(\psi x)^p$, we have

$$\frac{\phi x}{\psi x} = \frac{\phi f x}{(\psi x)^p} \quad \ldots \ldots \ldots \ldots (e)$$

Let us assume

$$y = \left\{\frac{\phi f x}{\psi f x}\right\}^{\frac{1}{p}} \left\{\frac{\phi f^2 x}{\psi f^2 x}\right\}^{\frac{1}{p}} \left\{\frac{\phi f^3 x}{\psi f^3 x}\right\}^{\frac{1}{p}} \left\{ \;\&\text{c.}\; \left\{\frac{\phi f^n x}{\psi f^n x}\right\}^{\frac{1}{p}} \right. \;*$$

$$y = \left\{\phi f x \left\{\frac{\phi f^2 x}{(\psi f x)^p}\right\}^{\frac{1}{p}} \left\{\frac{\phi f^3 x}{(\psi f^2 x)^p}\right\}^{\frac{1}{p}} \left\{ \;\&\text{c.}\; \left\{\frac{\phi f^n x}{(\psi f^{n-1} x)^p}\right\}^{\frac{1}{p}} \left\{\frac{1}{(\psi f^n x)^p}\right\}^{\frac{1}{p}} \right.$$

Multiplying by $(\phi f^{n+1} x)^{\frac{1}{p^{n+1}}}$, we have

$$y\,(\phi f^{n+1} x)^{\frac{1}{p^{n+1}}} = \left\{\phi f x \left\{\frac{\phi f^2 x}{(\psi f x)^p}\right\}^{\frac{1}{p}} \left\{\frac{\phi f^3 x}{(\psi f^2 x)^p}\right\}^{\frac{1}{p}} \left\{ \;\&\text{c.}\; \left\{\frac{\phi f^n x}{(\psi f^{n-1} x)^p}\right\}^{\frac{1}{p}} \left\{\frac{\phi f^{n+1} x}{(\psi f^n x)^p}\right\}^{\frac{1}{p}} \right.$$

Which becomes, by applying equation (e),

$$(\phi f^{n+1} x)^{\frac{1}{p^{n+1}}} y = \left\{\phi f x \left\{\frac{\phi f x}{\psi f x}\right\}^{\frac{1}{p}} \left\{\frac{\phi f^2 x}{\psi f^2 x}\right\}^{\frac{1}{p}} \left\{ \;\&\text{c.}\; \left\{\frac{\phi f^n x}{\psi f^n x}\right\}^{\frac{1}{p}} \right. = \left\{\phi f x \cdot y\right\}^{\frac{1}{p}}$$

Hence,
$$y^p \,(\phi f^{n+1} x)^{\frac{1}{p^n}} = y\,\phi f x$$

$$y^{p-1} = \frac{\phi f x}{(\phi f^{n+1} x)^{\frac{1}{p^n}}}$$

$$y = \left\{\frac{\phi f x}{(\phi f^{n+1} x)^{\frac{1}{p^n}}}\right\}^{\frac{1}{p-1}}$$

$$\left\{\frac{\phi f x}{(\phi f^{n+1} x)^{\frac{1}{p^n}}}\right\}^{\frac{1}{p-1}} = \left\{\frac{\phi f x}{\psi f x}\right\}^{\frac{1}{p}} \left\{\frac{\phi f^2 x}{\psi f^2 x}\right\}^{\frac{1}{p}} \left\{ \;\&\text{c.}\; \left\{\frac{\phi f^n x}{\psi f^n x}\right\}^{\frac{1}{p}} \right. \quad \ldots \ldots \ldots (f)$$

---

* Braces with an index as $\left(\frac{1}{p}\right)$ over them signify the $\frac{1}{p}$ power of all the following part of the expression.

Putting p for $\frac{1}{p}$, we find

$$\left\{\frac{(\phi f^{n+1}x)p^n}{\phi f x}\right\}^{\frac{p}{p-1}} = \left\{\frac{\phi f x}{\psi f x}\right\}^p \left\{\frac{\phi f^2 x}{\psi f^2 x}\right\}^p \left\{ \&\text{c.} \left\{\frac{\phi f^n x}{\psi f^n x}\right\}^p \cdots\cdots (g) \right.$$

And the equation determining the functions becomes,

$$\{\phi x\}^p \{\psi x\}^{p-1} = \{\phi f x\}^p$$

These equations in their present general form are sufficiently concise, but as they will become rather complex by the substitution of particular cases, I shall make use of the following notation to abbreviate them. P with an index above it placed before any function of n and other quantities signifies the continued product of that function, n being successively equal to 1. 2. and n; thus,

$$\overset{n}{P} \{\psi f^n x\} = \psi f x \cdot \psi f^2 x \cdot \psi f^3 x \cdots\cdots \psi f^n x$$

If the product is to be continued to infinity, I shall make use of Euler's method of denoting the limits of integrals.

In (b) assume

$$\psi x = 1 + x + x^2 + \&\text{c.} + x^a \cdots\cdots f x = x^{a+1}$$
$$\{1 + x + x^2 + \&\text{c.} + x^a\} \phi x = \phi x^{a+1}$$

or,
$$\frac{x^{a+1}-1}{x-1} \phi x = \phi x^{a+1}$$

Let, $\quad y_z = x$ and $y_{z+1} = x^{a+1}$ then,
$$y_z^{a+1} = y_{z+1}$$

By integrating $\quad y_z = y_0^{\overline{a+1}^z} = c^{\overline{a+1}^z} = x$

Let $\quad \phi x = \phi y_z = u_z$
$$\phi x^{a+1} = \phi y_{z+1} = u_{z+1}$$

then,
$$\frac{c^{\overline{a+1}^{z+1}}-1}{c^{\overline{a+1}^z}-1} u_z = z_{z+1}$$

$$(c^{\overline{a+1}^{z+1}}-1) u_z = (c^{\overline{a+1}^z}-1) u_{z+1}$$

But this equation is of the form

$$Q_{z+1} u_z = Q_z u_{z+1}$$

whose integral is

$$u_z = b Q_z$$

therefore,

$$u_z = \phi x = b \left(c^{\overline{a+1}}^{\frac{z}{}} - 1 \right) = (x - 1) b \ldots \ldots b = 1$$

Substituting these values of ψf and ϕ in (c), we have

$$\frac{x^{\overline{a+1}}^{n+1} - 1}{x^{a+1} - 1} = \overset{n}{P} \left\{ 1 + x^{1 \cdot \overline{a+1}}^{n} + x^{2 \cdot \overline{a+1}}^{n} + \&c. + x^{a \cdot \overline{a+1}}^{n} \right\} \ldots \ldots (1)$$

Let $a = 1$

$$\frac{x^{2}^{n+1} - 1}{x^2 - 1} = \overset{n}{P} \left\{ 1 + x^{2}^{n} \right\} = (1 + x^2)(1 + x^4)(1 + x^8) \ldots (1 + x^{2}^{n}) \ldots \ldots (2)$$

If $a = 2$

$$\frac{x^{3}^{n+1} - 1}{x^3 - 1} = \overset{n}{P} \left\{ 1 + x^{3}^{n} + x^{2 \cdot 3}^{n} \right\} = (1 + x^3 + x^6)(1 + x^9 + x^{18}) \ldots (1 + x^{3}^{n} + x^{2 \cdot 3}^{n}) \ldots (3)$$

In (1) for a put any even number as $2a$, and make x negative,

$$\frac{1 + x^{\overline{2a+1}}^{n+1}}{1 + x^{2a+1}} = \overset{n}{P} \left\{ 1 - x^{1 \cdot \overline{2a+1}}^{n} + x^{2 \cdot \overline{2a+1}}^{n} - \&c. + x^{2a \cdot \overline{2a+1}}^{n} \right\} \ldots \ldots \ldots (4)$$

Let $a = 1$

$$\frac{1 + x^{3}^{n+1}}{1 + x^3} = \overset{n}{P} \left\{ 1 - x^{3}^{n} + x^{2 \cdot 3}^{n} \right\} = (1 - x^3 + x^6)(1 - x^9 + x^{18}) \cdots (1 - x^{3}^{n} + x^{2 \cdot 3}^{n}) \ldots (5)$$

Assume $f x = x^2$, and

$$\psi x = 1 - x + x^2 - \&c. + x^{2a} = \frac{1 + x^{2a+1}}{1 + x}$$

$$x = y_z \qquad\qquad x^2 = y_{z+1}$$

$$y_z = c^{2}^{z}$$

Let $\qquad\qquad \phi y_z = u_z \qquad\qquad \phi y_{z+1} = u_{z+1}$

Equation (*b*) becomes

$$\frac{1 + c^{\overline{2a+1} \cdot 2^n}}{1 + c^z} u_z = u_{z+1}$$

$$u_z \left(1 + c^{\overline{2a+1} \cdot 2^z}\right) = u_{z+1} \left(1 + c^{2^z}\right)$$

In order to reduce this to the form of

$$Q_{z+1} u_z = Q_z u_{z+1}$$

multiply each side by $\left(1 - c^{2\overline{a+1} \cdot 2^z}\right) . \left(1 - c^{2^z}\right)$

$$u_z \left(1 + c^{2\overline{a+1} \cdot 2^z}\right) . \left(1 - c^{2\overline{a+1} \cdot 2^z}\right) . \left(1 - c^{2^z}\right) = u_{z+1} \left(1 - c^{2\overline{a+1} \cdot 2^z}\right) . \left(1 - c^{2^z}\right) . \left(1 + c^{2^z}\right)$$

$$u_z \left(1 - c^{\overline{2a+1} \cdot 2^{z+1}}\right) . \left(1 - c^{2^z}\right) = u_{z+1} \left(1 - c^{\overline{2a+1} \cdot 2^z}\right) . \left(1 - c^{2^{z+1}}\right)$$

$$u_z \times \frac{1 - c^{\overline{2a+1} \cdot 2^{z+1}}}{1 - c^{2^{z+1}}} = u_{z+1} \times \frac{1 - c^{\overline{2a+1} \cdot 2^z}}{1 - c^{2^z}}$$

Hence,

$$u_z = \phi x = \frac{1 - c^{\overline{2a+1} \cdot 2^z}}{1 - c^{2^z}} = \frac{1 - x^{\overline{2a+1}}}{1 - x}$$

Putting for f, ψ, and ϕ, their values in (*b*), we have

$$\frac{1 - x^{2\overline{a+1} \cdot 2^{n+1}}}{1 - x^{2 \cdot \overline{2a+1}}} \times \frac{1 - x^2}{1 - x^{2^{n+1}}} = \overset{n}{P} \left\{ 1 - x^{\cdot 2^n} + x^{2 \cdot 2^n} - \&c. + x^{2a \cdot 2^n} \right\} \quad \ldots \ldots (6)$$

Let $a = 1$,

$$\frac{1 - x^{3 \cdot 2^{n+1}}}{1 - x^{2 \cdot 3}} \times \frac{1 - x^2}{1 - x^{2^{n+1}}} = \frac{1 + x^{2^{n+1}} + x^{2^{n+2}}}{1 + x^2 + x^4} =$$

$$= \overset{n}{P} \left\{ 1 - x^{2^n} + x^{2^{n+1}} \right\} = (1 - x^2 + x^4)(1 - x^4 + x^8) \ldots (1 - x^{2^n} + x^{2^{n+1}}) \ldots \ldots (7)$$

Divide (5) by (3)

$$\frac{1 + x^{3 \cdot 2^{n+1}}}{1 - x^{3 \cdot 2^{n+1}}} \times \frac{1 - x^3}{1 + x^3} = P \left\{ \frac{1 - x^3 + x^{2 \cdot 3^n}}{1 + x^3 + x^{2 \cdot 3^n}} \right\} = \left(\frac{1 - x^3 + x^6}{1 + x^3 + x^9} \right) \ldots \left(\frac{1 - x^3 + x^{2 \cdot 3^n}}{1 + x^3 + x^{2 \cdot 3^n}} \right) \ldots (7)$$

In equation (1), for x put x^2, and let a be any even number as $2a$, then,

$$\frac{x^{\overline{2a+1}\cdot 2}^{\,n+1}-1}{x^{\overline{2a+1}\cdot 2}-1} = \overset{n}{P}\left\{1+x^{1\cdot\overline{2a+1}\cdot 2}+x^{2\cdot\overline{2a+1}\cdot 2}+\&c.+x^{2a\cdot\overline{2a+1}\cdot 2}\right\}$$

Divide both sides by

$$\overset{n}{P}\left\{x^{2a\cdot\overline{2a+1}}\right\} = x^{2a\left\{\overline{2a+1}+\overline{2a+1}+\&c.+\overline{2a+1}\right\}} = x^{\overline{2a+1}^{\,n+1}-\overline{2a+1}}$$

By properly arranging the right side of the equation, it becomes,

$$\frac{x^{\overline{2a+1}^{\,n+1}}-x^{-\overline{2a+1}^{\,n+1}}}{x^{\overline{2a+1}}-x^{-\overline{2a+1}}} =$$

$$= \overset{n}{P}\left\{1+(x^{2\cdot\overline{2a+1}}+x^{-2\cdot\overline{2a+1}})+(x^{4\cdot\overline{2a+1}}+x^{-4\cdot\overline{2a+1}})+\&c.+(x^{2a\cdot\overline{2a+1}}+x^{-2a\cdot\overline{2a+1}})\right\}$$

For $\qquad\qquad x^{+1}+x^{-1}\qquad$ put $\qquad 2\cos.\theta$

This substitution gives, $\qquad \dfrac{\sin.\overline{2a+1}^{\,n+1}.\theta}{\sin.\overline{2a+1}.\theta} =$

$$= \overset{n}{P}\left\{1+2\cos.2.\overline{2a+1}\,\theta+2\cos.4.\overline{2a+1}\,\theta+\&c.+2\cos.2a.\overline{2a+1}\,\theta\right\}\ \ldots\ldots(8)$$

For a in (1), put any odd number as $2a-1$, and substituting x^2 for x,

$$\frac{x^{(2a)\cdot 2}^{\,n+1}-1}{x^{(2a)\cdot 2}-1} = \overset{n}{P}\left\{1+x^{1\cdot(2a)\cdot 2}+x^{2\cdot(2a)\cdot 2}+\&c.+x^{\overline{2a-1}\cdot(2a)\cdot 2}\right\}$$

Divide each side by

$$\overset{n}{P}\left\{x^{\overline{2a-1}\cdot(2a)}\right\} = x^{\overline{2a-1}\left\{(2a)^1+(2a)^2+\&c.(2a)^n\right\}} = x^{(2a)^{n+1}-2a}$$

then,

$$\frac{x^{(2a)^{n+1}}-x^{-(2a)^{n+1}}}{x^{2a}-x^{-2a}} =$$

$$= \overset{n}{P}\left\{(x^{1\cdot(2a)}+x^{-1\cdot(2a)})+(x^{3\cdot(2a)}+x^{-3\cdot(2a)})+\&c.+(x^{\overline{2a-1}\cdot(2a)}+x^{-\overline{2a+1}\cdot(2a)})\right\}$$

Using the substitution

$$x^{+1} + x^{-1} = 2 \cos. \theta$$

$$\frac{\sin. (2a)^{n+1} \theta}{\sin. 2a\; \theta} = P\left\{\cos. 1\,(2a)^n\theta + \cos. 3\,(2a)^n\theta + \&c. + \cos. \overline{2a-1}\,(2a)^n\theta\right\} \times \overset{n}{P}\{2\}$$

But
$$\overset{n}{P}\{2\} = 2^n$$

$$\frac{1}{2^n} \cdot \frac{\sin. (2a)^{n+1} \theta}{\sin. 2a\; \theta} = P\left\{\cos. 1\,(2a)^n\theta + \cos. 3\,(2a)^n\theta + \&c. + \cos. \overline{2a-1}\,(2a)^n\theta.\right\}..\,(9)$$

If $a = 1$, it becomes the well known expression of Euler

$$\frac{1}{2^n} \cdot \frac{\sin. 2^{n+1}\theta}{\sin. 2\theta} = \cos. 2\theta . \cos. 2^2\theta \ldots\ldots \cos. 2^n\theta \ldots\ldots\ldots (1,1)$$

Let $a = 2$

$$\frac{1}{2^n} \cdot \frac{\sin. 4^{n+1}\theta}{\sin. 4\theta} = P\left\{\cos. 1.4^n\theta + \cos. 3.4^n\theta\right\} \ldots\ldots\ldots\ldots (1,2)$$

If the same operations be performed on (4) and (6) that have been made use of on (1), the results will be

$$\frac{\cos. \overline{2a+1}^{\,n+1}\theta}{\cos. \overline{2a+1}\;\theta} = \overset{n}{P}\left\{\pm(1 - 2\cos. 2.\overline{2a+1}^{\,n}\theta + \&c. + 2\cos. 2a.\overline{2a+1}^{\,n}\theta)\right\}$$

and

$$\frac{\sin. 2.\overline{2a+1}^{\,n+1}.\theta}{\sin. 2\;\overline{2a+1}\;\theta} \cdot \frac{\sin. 2\theta}{\sin. 2^n\theta} =$$

$$= \overset{n}{P}\left\{\pm(1 - 2\cos. 2.2^n\theta + 2\cos. 4.2^n\theta - \&c. + 2\cos. 2a.2^n\theta)\right\}$$

plus or minus being used as a is an even or odd number.

To obviate the ambiguity of the signs, it will be better to put for a, $2a$, and $2a - 1$, then,

$$\frac{\cos. \overline{4a+1}^{\,n+1}.\theta}{\cos. \overline{4a+1}\;\theta} =$$

$$= \overset{n}{P}\left\{1 - 2\cos. 2.\overline{4a+1}^{\,n}.\theta + 2\cos. 4.\overline{4a+1}^{\,n}\theta - \&c. + 2\cos. 4a.\overline{4a+1}^{\,n}\theta\right\} \ldots (1,3)$$

$$\frac{\cos.\ \overline{4a-1}\ \theta}{\cos.\ \overline{4a-1}\ \theta}^{n+1} \ =$$

$$= \overset{n}{P}\left\{2\ \cos.\overline{4a-2}.\overline{4a-1}^{n}\theta - 2\cos.\overline{4a-4}.\overline{4a-1}^{n}\theta + \&c. + 2\cos.2\overline{4a-1}^{n}\theta - 1\right\}..\,(1,4)$$

$$\frac{\sin.\ 2\ \overline{4a+1}\ \theta}{\sin.\ 2\ \overline{4a+1}\ \theta}^{n+1} \cdot \frac{\sin.\ 2\ \theta}{\sin.\ 2^{+1}\theta} \ =$$

$$= \overset{n}{P}\left\{1 - 2\cos.2.\overset{n}{2}\,\theta + 2\cos.4.\overset{n}{2}\,\theta - \&c. + 2\cos.4a.\overset{n}{2}\,\theta\right\}\dots\dots\,(1,5)$$

$$\frac{\sin.\ 2\ \overline{4a-1}\ \theta}{\sin.\ 2.\overline{4a-1}\ \theta}^{n+1} \cdot \frac{\sin.\ 2\,\theta}{\sin.\ 2^{n+1}\theta} \ =$$

$$= \overset{n}{P}\left\{2\cos.\overline{4a-2}.\overset{n}{2}\,\theta - 2\cos.\overline{4a-4}.\overset{n}{2}\,\theta + \&c. + 2\cos.2.\overset{n}{2}\,\theta - 1\right\}\dots\,(1,6)$$

In (8) make $a = 1$, and also make $a = 1$ in (1, 4)   Dividing the last result by the first.

$$\frac{\cos.3\ \theta}{\cos.3\ \theta}^{n+1} \cdot \frac{\sin.3\ \theta}{\sin.3^{n+1}\theta} = \frac{\tan.3\ \theta}{\tan.3^{n+1}\theta} = \overset{n}{P}\left\{\frac{2\cos.2.3\ \theta-1}{2\cos.2.3^n\theta+1}\right\} = \overset{n}{P}\left\{\frac{2-\sec.2.3\ \theta}{2+\sec.2.3^n\theta}\right\}..\,(1,7)$$

Or,

$$\frac{\tan.\ \theta}{\tan.3^{n+1}\theta} = \left\{\frac{2-\sec.2.3\ \theta}{2+\sec.2.3\ \theta}\right\} \cdot \left\{\frac{2-\sec.2.3^2\theta}{2+\sec.2.3^2\theta}\right\}\dots\left\{\frac{2-\sec.2.3^n\theta}{2+\sec.2.3^n\theta}\right\}$$

In (1, 6) make $a = 1$, and in (1, 1), put 2θ for θ;   then,

$$\frac{\sin.3.2^{n+1}\theta}{\sin.2^{n+1}\theta} \cdot \frac{\sin.2\,\theta}{\sin.3.2\theta} = \overset{n}{P}\left\{2\cos.2.2^n\theta - 1\right\}\dots\dots\,(1,8)$$

And, $$\frac{1}{2^n} \cdot \frac{\sin.2.2^{n+1}\theta}{\sin.2.2\theta} = \overset{n}{P}\left\{\cos.2.2^n\theta\right\}$$

The first divided by the second produces

$$\frac{2^n\sin.3.2^{n+1}\theta}{\sin.3.2\theta} \cdot \frac{\sin.2\,\theta\,\sin.2.2\,\theta}{\sin.2^{n+1}\theta\,\sin.2.2^{n+1}\theta} = \overset{n}{P}\left\{2-\sec.2.2^n\theta\right\}\dots\,(1,9)$$

If in theorems (8) (9) (1, 3) (1, 4) (1, 5) (1, 6) for θ be substituted in each respectively,

$$\frac{\theta}{(2a+1)^{n+1}}, \quad \frac{\theta}{(2a)^{n+1}}, \quad \frac{\theta}{(4a+1)^{n+1}}, \quad \frac{\theta}{(4a-1)^{n+1}}, \quad \frac{\theta}{2^{n+1}}, \quad \frac{\theta}{2^{n+1}}$$

The following are the results:

$$\frac{\sin.\,\theta}{\sin.\,\dfrac{\theta}{(2a+1)^n}} = \overset{n}{P} \left\{ 1 + 2\cos.\,\frac{2\theta}{(2a+1)^n} + 2\cos.\,\frac{4\theta}{(2a+1)^n} + \&c. + 2\cos.\,\frac{2a\theta}{(2a+1)^n} \right\} \;\ldots (2,1)$$

$$\frac{1}{2^n} \cdot \frac{\sin.\,\theta}{\sin.\,\dfrac{\theta}{(2a)^n}} = \overset{n}{P} \left\{ \cos.\,\frac{1.\theta}{(2a)^n} + \cos.\,\frac{3.\theta}{(2a)^n} + \&c. + \cos.\,\frac{\overline{2a-1}\,\theta}{(2a)^n} \right\} \;\ldots\ldots (2,2)$$

$$\frac{\cos.\,\theta}{\cos.\,\dfrac{\theta}{(4a+1)^n}} = \overset{n}{P} \left\{ 1 - 2\cos.\,\frac{2\theta}{(4a+1)^n} + 2\cos.\,\frac{4\theta}{(4a+1)^n} - \&c. + 2\cos.\,\frac{4a\theta}{(4a+1)^n} \right\} \;\ldots (2,3)$$

$$\frac{\cos.\,\theta}{\cos.\,\dfrac{\theta}{(4a-1)^n}} = \overset{n}{P} \left\{ 2\cos.\,\frac{\overline{4a-2}\,\theta}{(4a-1)^n} - 2\cos.\,\frac{\overline{4a-4}\,\theta}{(4a-1)^n} + \&c. + 2\cos.\,\frac{2\theta}{(4a-1)^n} - 1 \right\} \;\ldots (2,4)$$

$$\frac{\sin.\,\overline{4a+1}\,\theta}{\sin.\,\dfrac{4a+1}{2^n}\theta} \cdot \frac{\sin.\,\dfrac{\theta}{2^n}}{\sin.\,\theta} = \overset{n}{P} \left\{ 1 - 2\cos.\,\frac{2\theta}{2^n} + 2\cos.\,\frac{4\theta}{2^n} - \&c. + 2\cos.\,\frac{4a\theta}{2^n} \right\} \;\ldots (2,5)$$

$$\frac{\sin.\,\overline{4a-1}\,\theta}{\sin.\,\dfrac{4a-1}{2^n}\theta} \cdot \frac{\sin.\,\dfrac{\theta}{2^n}}{\sin.\,\theta} = \overset{n}{P} \left\{ 2\cos.\,\frac{\overline{4a-2}\,\theta}{2^n} - 2\cos.\,\frac{\overline{4a-4}\,\theta}{2^n} + \&c. + 2\cos.\,\frac{2\theta}{2^n} - 1 \right\} \;\ldots (2,6)$$

Other theorems nearly similar may be thus derived. In (1), for x put vx, and $\dfrac{x}{v}$; multiply together the results; and in the left side of the equation put for $v^{+1} + v^{-1}$ its value $2\cos.\,\theta$; then,

$$\frac{x^{2(a+1)^{n+1}} - 2x^{(a+1)^{n+1}}\cos.\,\overline{a+1}^{\,n+1}\theta + 1}{x^{2(a+1)} - 2x^{a+1}\cos.\,\overline{a+1}\,\theta + 1} =$$

$$= P \left\{ 1 + (xv)^{1.\overline{a+1}^{\,n}} + (xv)^{2.\overline{a+1}^{\,n}} + \&c. + (xv)^{a.\overline{a+1}^{\,n}} \right\}$$

$$\times \; P \left\{ 1 + \left(\frac{x}{v}\right)^{1.\overline{a+1}^{\,n}} + \left(\frac{x}{v}\right)^{2.\overline{a+1}^{\,n}} + \&c. + \left(\frac{x}{v}\right)^{a.\overline{a+1}^{\,n}} \right\}$$

c

In order to reduce the right side of the equation to a series of multiple arcs, let us consider the product of two functions of this form;

$$f(xv) = 1 + (xv) + (xv)^2 + \&c. + (xv)^a$$

$$f\left(\frac{x}{v}\right) = 1 + \left(\frac{x}{v}\right) + \left(\frac{x}{v}\right)^2 + \&c. + \left(\frac{x}{v}\right)^a$$

Their product is of the form

$$f(xv) \times f\left(\frac{x}{v}\right) =$$

$$(1+x^2+\&c.x^{2a}) + x(1+x^2+\&c.x^{2a-2})(v^{+1}+v^{-1}) + x^2(1+x^2+\&c.+x^{2a-4})(v^{+2}+v^{-2}) + \&c. + x^a(v^{+a}+v^{-a})$$

$$= \{1-x^2\}^{-1}.\{ (1-x^{2a+2}) + x(1-x^{2a})(v^{+1}+v^{-1}) + x^2(1-x^{2a-2})(v^{+2}+v^{-2}) + \&c.x^a(v^{+a}+v^{-a})\}$$

Putting $v^{+1}+v^{-1} = 2\cos. \theta$, and applying this to the preceding expression,

$$\frac{x^{2.a+1} - 2x^{\overline{a+1}\frac{n+1}{a+1}}\cos. \overline{a+1}^{\frac{n+1}{}}\theta + 1}{x^{2.a+1} - 2x^{a+1}\cos. \overline{a+1}\theta + 1} =$$

$$= (1-x^2)^{-n} \overset{n}{P} \{(1-x^{\overline{2a+2}.\overline{a+1}}) + 2(1-x^{\overline{2a.a+1}})x^{\overline{1.a+1}}\cos. 1.\overline{a+1}^{\,n}\theta + \&c. + 2x^{\overline{a.a+1}}\cos.a.\overline{a+1}^{\,n}\theta\}$$

In the case of x equal to unity, this expression becomes

$$\frac{1 - \cos. \overline{a+1}^{\frac{n+1}{}}\theta}{1 - \cos. \overline{a+1}\theta} =$$

$$P\{(a+1) + 2(a)\cos. 1.\overline{a+1}^{\,n}\theta + 2(a-1)\cos. 2\overline{a+1}^{\,n}\theta + \&c. + 2(1)\cos.a.\overline{a+1}^{\,n}\theta\}..(2,6)$$

If x do not equal unity, and if $a=1$;

$$\frac{x^{2.2} - 2x^2{}^{\frac{n+1}{}}\cos. 2^{n+1}\theta + 1}{x^{2.2} - 2x^2\cos. 2\theta + 1} = P\{1 + 2x^{2\,n}\cos. 2^n\theta + x^{2\,{}^{n+1}}\}......(2,7)$$

If the same operations are performed on (4) and (6), we shall obtain the following results;

$$\frac{1 + \cos. \overline{2a+1}.{}^{\frac{n+1}{}}\theta}{1 + \cos. \overline{2a+1}.\theta} =$$

$$= \overset{n}{P}\{(2a+1) - 2(2a)\cos. 1.\overline{2a+1}^{\,n}\theta + 2(2a-1)\cos. 2\overline{2a+1}^{\,n}\theta - \&c. + 2(1)\cos. 2a.\overline{2a+1}^{\,n}\theta\}..(2,8)$$

$$\frac{1-\cos.\,\overline{2\,a+1}\,.\,2^{n+1}\,\theta}{1-\cos.\,\overline{2\,a+1}\,.\,2\,\theta}\cdot\frac{1-\cos.\,2\,\theta}{1-\cos.\,2^{n+1}\theta}=\left\{\frac{\sin.\,\overline{2\,a+1}.2^{n}\theta}{\sin.\,\overline{2\,a+1}.\theta}\cdot\frac{\sin.\,\theta}{\sin.\,2^{n}\theta}\right\}^{2}=$$

$$=\overset{n}{P}\left\{(2a+1)-2(2a)\cos.1.2^{n}\theta+2(2a-1)\cos.2.2^{n}\theta-\&c.+2(1)\cos.2a.2^{n}\theta\right\}\ ..\,(2,9)$$

We will now give some examples of the use of theorem (f), the chief difficulty consists in finding functions which satisfy the equation (d), for, if the two functions ψ and f be assumed, the third must be determined by the method of La Place, and this generally leads to equations of finite differences whose solutions are above the powers of Analysis in its present state.

In one case however (namely when $p=2$), the solutions we have obtained for equation (b), are also solutions of (d), for in this case (b) and (d) are identical.

We have already determined, that if

$$\psi x=1+x+x^{2}+\&c.\,x^{a},\quad\text{and }fx=x^{a+1}$$

that $$\phi x=x-1\,;$$

hence, substituting these values in (f),

$$\frac{x^{a+1}-1}{\left\{\overline{x^{a+1}}^{n+1}-1\right\}^{\frac{1}{2}n}}=\overset{n}{P}\left\{\frac{x^{\overline{a+1}^{n}}-1}{1+x^{1.\overline{a+1}^{n}}+x^{2.\overline{a+1}^{n}}+\&c.+x^{a.\overline{a+1}^{n}}}\right\}^{\frac{1}{2}n}\ ......\,(3,1)$$

If $a=1$,

$$\frac{x^{2}-1}{\left\{\overline{x^{2}}^{n+1}-1\right\}^{\frac{1}{2}n}}=\overset{n}{P}\left\{\frac{x^{2^{n}}-1}{x^{2^{n}}+1}\right\}^{\frac{1}{2}n}=\left\{\frac{x^{2}-1}{x^{2}+1}\right\}^{\frac{1}{2}}\left\{\frac{x^{4}-1}{x^{4}+1}\right\}^{\frac{1}{2}}\left\{\frac{x^{8}-1}{x^{8}+1}\right\}^{\frac{1}{2}}\ \&c.\ \left\{\frac{x^{2^{n}}-1}{x^{2^{n}}+1}\right\}^{\frac{1}{2}}\ .....\,(3,2)$$

put xv, and xv^{-1} for x, multiply the results

$$\frac{x^{2}v^{2}-1}{\left\{(xv)^{2^{n+1}}-1\right\}^{\frac{1}{2}n}}\times\frac{x^{2}v^{-2}-1}{\left\{(xv^{-1})^{2^{n+1}}-1\right\}^{\frac{1}{2}n}}=\overset{n}{P}\left\{\frac{x^{2^{n+1}}-x^{2^{n}}(v^{2}+v^{-2})+1}{x^{2^{n+1}}+x^{2^{n}}(v^{2}+v^{-2})+1}\right\}^{\frac{1}{2}n}$$

for $v+v^{-1}$ put $2\cos.\theta$

$$\frac{x^{4}-2x^{2}\cos.2\,\theta+1}{\left\{x^{2^{n+2}}-2x^{2^{n+1}}\cos.2^{n+1}\theta+1\right\}^{\frac{1}{2}n}}=\overset{n}{P}\left\{\frac{x^{2^{n+1}}-2x^{2^{n}}\cos.2^{n}\theta+1}{x^{2^{n+1}}+2x^{2^{n}}\cos.2^{n}\theta+1}\right\}^{\frac{1}{2}n}\ ....\,(3,3)$$

If $x = 1$, and for θ be substituted 2θ, and the necessary reductions be made,

$$\frac{2\sin.2\theta}{\left(2\sin.2^{n+1}\theta\right)^{\frac{1}{2^n}}} = \overset{n}{P}\left\{\tan.2^n\theta\right\}^{\frac{1}{2^n}} = \left(\tan.2\theta\right)^{\frac{1}{2}}.\left(\tan.4\theta\right)^{\frac{1}{4}}\cdots\left(\tan.2^n\theta\right)^{\frac{1}{2^n}}..(3,4)$$

For θ substitute $\dfrac{\theta}{2^{n+1}}$

$$\frac{2\sin.\dfrac{\theta}{2^n}}{\left(2\sin.\theta\right)^{\frac{1}{2^n}}} = \left(\tan.\frac{\theta}{2^n}\right)^{\frac{1}{2}}.\left(\tan.\frac{\theta}{2^{n-1}}\right)^{\frac{1}{4}}\cdots\left(\tan.\frac{\theta}{2}\right)^{\frac{1}{2^n}}$$

Raise both sides to the power 2^{n+1}, then,

$$\frac{\left(2\sin.\dfrac{\theta}{2^n}\right)^{2^{n+1}}}{\left(2\sin.\theta\right)^2} = \overset{n}{P}\left\{\tan.\frac{\theta}{2^n}\right\}^{2^n} = \left(\tan.\frac{\theta}{2}\right)^2.\left(\tan.\frac{\theta}{4}\right)^4\cdots\left(\tan.\frac{\theta}{2^n}\right)^{2^n}..(3,5)$$

Assume

$$fx = x^{a+1} \qquad \phi x = 1-x$$

$$(\psi x)^{p-1} = \frac{\phi f x}{\phi x} = \frac{1-x^{a+1}}{1-x} = 1+x+x^2+\&c.+x^a$$

$$\psi x = {}^{p-1}\!\sqrt{1+x+x^2+\&c.+x^a}$$

and by substituting the values in (f),

$$\left\{\frac{1-x^{a+1}}{\left\{1-x^{\overline{a+1}}\right\}^{\frac{1}{p^n}}}\right\}^{\frac{1}{p-1}} = \overset{n}{P}\left\{\frac{1-x^{\overline{a+1}}^n}{{}^{p-1}\!\sqrt{1+x^{1.\overline{a+1}^n}+x^{2.\overline{a+1}^n}+\&c.+x^{a.\overline{a+1}^n}}}\right\}^{\frac{1}{p^n}}\ldots(3,6)$$

For a put $2a$, and for x put x^2. Divide both sides by $x^{\frac{1}{p-1}\left(\overline{2a+1}-\frac{\overline{2a+1}}{p^n}\right)^{n+1}}$ and substituting for $x^{+1}+x^{-1}$ its value $2\cos.\theta$, and for $x^{+1}-x^{-1}$ its value $+2\sqrt{-1}\sin.\theta$, some power of $\sqrt{-1}$ will enter both sides of the equation, and they must be divided by it, but the $\sqrt{-1}$ is on both sides of the equation multiplied by 2; and therefore the equation is also divisible by the same power of 2; having performed these operations, the result will be

$$\left\{\frac{\sin.\overline{2a+1}\,\theta}{\left\{\sin.\overline{2a+1}^{n+1}\theta\right\}^{\frac{1}{p^n}}}\right\}^{\frac{1}{p-1}} =$$

$$= \overset{n}{P}\left\{\frac{\sin.\overline{2a+1}^n\theta}{{}^{p-1}\!\sqrt{1+2\cos.2.\overline{2a+1}^n\theta+2\cos.4.\overline{2a+1}^n\theta+\&c.+2\cos.2a.\overline{2a+1}^n\theta}}\right\}^{\frac{1}{p^n}}..(3,7)$$

If we had began by putting $2a$ for a, and $-x$ for x, the result would have been

$$\left\{ \frac{\cos.\overline{2a+1}\,\theta}{\left\{\cos.\overline{2a+1}^{n+1}\theta\right\}^{\frac{1}{p^n}}} \right\}^{\frac{1}{p-1}} =$$

$$= \overset{n}{P}\left\{ \frac{\cos.\overline{2a+1}^{\,n}\theta}{\sqrt[p-1]{1-2\cos.2.\overline{2a+1}^{\,n}\theta+2\cos.4\,\overline{2a+1}^{\,n}\theta-\&c.+2\cos.2a.\overline{2a+1}^{\,n}\theta}} \right\}^{\frac{1}{p^n}} . (3,8)$$

Dividing the first of these expressions by the second,

$$\left\{ \frac{\tan.\overline{2a+1}\,\theta}{\left\{\tan.\overline{2a+1}^{n+1}\theta\right\}^{\frac{1}{p^n}}} \right\}^{\frac{1}{p-1}} =$$

$$= \overset{n}{P}\left\{ \frac{\sin.\overline{2a+1}^{\,n}\theta}{\cos.\overline{2a+1}^{\,n}\theta}\left\{\frac{1-2\cos.2.\overline{2a+1}^{\,n}\theta+2\cos.4.\overline{2a+1}^{\,n}\theta-\&c.+2\cos.2a.\overline{2a+1}^{\,n}\theta}{1+2\cos.2.\overline{2a+1}^{\,n}\theta+2\cos.4.\overline{2a+1}^{\,n}\theta-\&c.+2\cos.2a.\overline{2a+1}^{\,n}\theta}\right\}^{\frac{1}{p-1}} \right\}^{\frac{1}{p^n}} .. (3,9)$$

If $a = 1$, $p = 2$,

$$\frac{\tan.3\,\theta}{\left\{\tan.3^{n+1}\theta\right\}^{\frac{1}{2^n}}} = \overset{n}{P}\left\{\tan.3^n\theta . \frac{\sec.2.3^n\theta-2}{\sec.2.3^n\theta+2}\right\}^{\frac{1}{2^n}} .... (4,1)$$

In $(3,6)$ for x, substitute successively xv and xv^{-1}; multiplying the results, and making

$$x = 1, \text{ and } v^{+1}+v^{-1} = 2\cos.2\theta,$$

we have

$$\left\{ \frac{\sin.\overline{a+1}\,\theta}{\left\{\sin.\overline{a+1}^{n+1}\theta\right\}^{\frac{1}{p^n}}} \right\}^{\frac{2}{p-1}} =$$

$$= \overset{n}{P}\left\{ \frac{(\sin.\overline{a+1}^{\,n}\theta)^2}{\sqrt[p-1]{(a+1)+2(a)\cos.2.\overline{a+1}^{\,n}\theta+2(a-1)\cos.4.\overline{a+1}^{\,n}\theta+\&c.+2(1)\cos.2a.\overline{a+1}^{\,n}\theta}} \right\}^{\frac{1}{p^n}} ..(4,2)$$

If before the preceding substitution x had been made negative, and $2a$ put for a, we should have found

D

$$\left\{ \frac{\cos. \overline{2a+1}\,\theta}{\left\{\cos. \overline{2a+1}^{\,n+1}\theta\right\}^{\frac{1}{p^n}}} \right\}^{\frac{1}{p-1}} =$$

$$= \overset{n}{P}\left\{ \frac{\cos. \overline{2a+1}^{\,n}\theta}{\sqrt{(2a+1)-2(2a)\cos.2.\overline{2a+1}^{\,n}\theta+2(2a-1)\cos.4.\overline{2a+1}^{\,n}\theta-\&c.+2(1)\cos.4a.\overline{2a+1}^{\,n}\theta}} \right\}^{\frac{1}{p^n}} \!\!\!\!.\,.\,(4,3)$$

In $(4, 2)$ make $p = \dfrac{3}{2}$, and for a put $2a$,

$$\left\{ \frac{\sin. \overline{2a+1}\,\theta}{\left\{\sin. \overline{2a+1}^{\,n+1}\theta\right\}^{(\frac{2}{3})^n}} \right\}^{2} =$$

$$= \overset{n}{P}\left\{ \frac{\sin. \overline{2a+1}^{\,n}\theta}{(2a+1)+2(2a)\cos.2.\overline{2a+1}^{\,n}\theta+2(2a-1)\cos.4.\overline{2a+1}^{\,n}\theta+\&c.+2(1)\cos.4a.\overline{2a+1}^{\,n}\theta} \right\}^{(\frac{2}{3})^n} \!\!\!\!.\,.\,(4,5)$$

In $(4, 3)$ make $p = \dfrac{3}{2}$,

$$\left\{ \frac{\cos. \overline{2a+1}\,\theta}{\left\{\cos. \overline{2a+1}^{\,n+1}\theta\right\}^{(\frac{2}{3})^n}} \right\}^{2} =$$

$$= \overset{n}{P}\left\{ \frac{\cos. \overline{2a+1}^{\,n}\theta}{(2a+1)-2(2a)\cos.2.\overline{2a+1}^{\,n}\theta+2(2a-1)\cos.4.\overline{2a+1}^{\,n}\theta-\&c.+2(1)\cos.4a.\overline{2a+1}^{\,n}\theta} \right\}^{(\frac{2}{3})^n} \!\!\!\!.\,.\,(4,6)$$

Dividing the first by the last,

$$\left\{ \frac{\tan. \overline{2a+1}^{\,n}\theta}{\left\{\tan. \overline{2a+1}^{\,n+1}\theta\right\}^{(\frac{2}{3})^n}} \right\}^{2} =$$

$$\overset{n}{P}\left\{ \frac{\sin.\overline{2a+1}^{\,n}\theta}{\cos.\overline{2a+1}^{\,n}\theta} . \frac{(2a+1)-2(2a)\cos.2.\overline{2a+1}^{\,n}\theta+2(2a-1)\cos.4.\overline{2a+1}^{\,n}\theta-\&c.+2(1)\cos.4a.\overline{2a+1}^{\,n}\theta}{(2a+1)+2(2a)\cos.2.\overline{2a+1}^{\,n}\theta+2(2a-1)\cos.4.\overline{2a+1}^{\,n}\theta+\&c.+2(1)\cos.4a.\overline{2a+1}^{\,n}\theta} \right\}^{(\frac{2}{3})^n} \!\!\!\!.\,(4,7)$$

In $(4, 2)$ let $p = \dfrac{3}{2}$, $a = 1$

$$\left\{ \frac{\sin. 2\,\theta}{\left\{\sin. 2^{n+1}\theta\right\}}(\tfrac{2}{3})^n \right\} = \overset{n}{P}\left\{ \frac{\sin. 2^n\theta}{2(1+\cos. 2^{n+1}\theta)} \right\}^{(\tfrac{2}{3})^n} =$$

$$= \overset{n}{P}\left\{ \frac{\sin. 2^n\theta}{(\cos. 2^n\theta)^2} \right\}^{(\tfrac{2}{3})^n} \times \overset{n}{P}\left\{ \frac{1}{4} \right\}^{(\tfrac{2}{3})^n}$$

but, $\quad \overset{n}{P}\left\{\frac{1}{4}\right\}^{(\tfrac{2}{3})^n} = \left\{\frac{1}{4}\right\}^{\tfrac{2}{3}+\tfrac{4}{9}+\&c.+\tfrac{2^n}{3^n}} = \left\{\frac{1}{4}\right\}^{\frac{(\tfrac{2}{3})^{n+1}-\tfrac{2}{3}}{-\tfrac{1}{3}}} = \left\{\frac{1}{4}\right\}^{2(1-(\tfrac{2}{3})^n)} = \left\{\frac{4}{4}^{(\tfrac{2}{3})^n}\right\}^2$

therefore, multiplying both sides by $\left\{\frac{4}{4}^{(\tfrac{2}{3})^n}\right\}^2$

$$\left(\frac{4\sin. 2\,\theta}{\left\{4\sin. 2^{n+1}\theta\right\}}(\tfrac{2}{3})^n \right)^2 = \overset{n}{P}\left\{ \tan. 2^n\theta\ \sec. 2^n\theta \right\}^{(\tfrac{2}{3})^n} =$$

$$= \left(\tan. 2\theta\sec. 2\theta\right)^{\tfrac{2}{3}}.\left(\tan. 4\theta.\sec. 4\theta\right)^{\tfrac{4}{9}}.\left(\tan. 8\theta\sec. 8\theta\right)^{\tfrac{8}{27}}...\left(\tan. 2^n\theta.\sec. 2^n\theta\right)^{(\tfrac{2}{3})^n}.(4,8)$$

For θ put $\dfrac{\theta}{2^{n+1}}$, and raise both sides to the power $\left(\dfrac{3}{2}\right)^{n+1}$, then we find

$$\left\{ \frac{\left\{4\sin. \dfrac{\theta}{2^n}\right\}^{(\tfrac{3}{2})^{n+1}}}{\left\{4\sin. \theta\right\}^{\tfrac{3}{2}}} \right\}^2 = \overset{n}{P}\left\{ \tan. \frac{\theta}{2^n}\ \sec. \frac{\theta}{2^n} \right\}^{(\tfrac{3}{2})^n} =$$

$$= \left(\tan. \frac{\theta}{2}\sec. \frac{\theta}{2}\right)^{\tfrac{3}{2}}.\left(\tan. \frac{\theta}{4}\sec. \frac{\theta}{4}\right)^{\tfrac{9}{4}}.\left(\tan. \frac{\theta}{8}\sec. \frac{\theta}{8}\right)^{\tfrac{27}{8}}...\left(\tan. \frac{\theta}{2^n}\sec. \frac{\theta}{2^n}\right)^{(\tfrac{3}{2})^n}.(4,9)$$

In $(3,6)$ for a substitute $2a-1$, put x^2 for x, and

$$2\cos. \theta = x^{+1}+x^{-1}$$

$$\left\{ \frac{\sin. 2a\,\theta}{\left\{\sin. (2a)^{n+1}\theta\right\}^{\tfrac{1}{p^n}}} \right\}^{\tfrac{1}{p-1}} =$$

$$= \overset{n}{P}\left\{ \frac{\sin. (2a)^n\theta}{\sqrt[p-1]{\cos. 1.(2a)^n\theta+\cos. 3.(2a)^n\theta+\&c.+\cos. \overline{2a-1}.(2a)^n\theta}} \right\}^{\tfrac{1}{p^n}} \times \overset{n}{P}\left\{ \frac{1}{(2)^{\tfrac{1}{p-1}}} \right\}^{\tfrac{1}{p^n}}$$

But

$$\overset{n}{P}\left\{ \frac{1}{(2)^{\tfrac{1}{p-1}}} \right\}^{\tfrac{1}{p^n}} = \left\{ \frac{1}{2^{\tfrac{1}{p-1}}} \right\}^{\tfrac{1}{p}+\tfrac{1}{p^2}+\&c.\tfrac{1}{p^n}} = \left\{ \frac{1}{2^{\tfrac{1}{p-1}}} \right\}^{\frac{\tfrac{1}{p^{n+1}}-\tfrac{1}{p}}{\tfrac{1}{p}-1}} = \left\{ \frac{1}{2^{\tfrac{1}{p-1}}} \right\}^{\frac{1-\tfrac{1}{p^n}}{p-1}} = 2^{-\frac{1}{(p-1)_2}(1-\tfrac{1}{p^n})}$$

then,

$$\left\{\frac{2^{\frac{1}{p-1}}\sin.2a\,\theta}{\left\{2^{\frac{1}{p-1}}\sin.(2a)^{n+1}\,\theta\right\}^{\frac{1}{p^n}}}\right\}^{\frac{1}{p-1}} =$$

$$= \overset{n}{P}\left\{\frac{\sin.(2a)^n\theta}{\overset{p-1}{\sqrt{\cos.1\,(2a)^n\,\theta+\cos.3.(2a)^n\,\theta+\&c.+\cos.\overline{2a-1}.(2a)^n\,\theta}}}\right\}^{\frac{1}{p^n}}\ \ldots (5,0)$$

In this expression make $a = 2$, $p = 2$,

$$\cos.4^n\theta+\cos.3.4^n\theta = \cos.4^n\theta+2\cos.4^n\theta\cos.2.4^n\theta-\cos.4^n\theta = 2\cos.4^n\theta\cos.2.4^n\theta$$

then,

$$\frac{4\sin.4\,\theta}{\left\{4\sin.4^{n+1}\theta\right\}^{\frac{1}{2^n}}} = \overset{n}{P}\left\{\tan.4^n\theta\sec.2.4^n\theta\right\}^{\frac{1}{2^n}} =$$

$$= \left\{\tan.4\theta\sec.8\theta\right\}^{\frac{1}{2}}.\left\{\tan.16\theta\sec.32\theta\right\}^{\frac{1}{4}}..\left\{\tan.4^n\theta\sec.2.4^n\theta\right\}^{\frac{1}{2^n}}\ \ldots (5,1)$$

For θ substitute $\dfrac{\theta}{4^{n+1}}$, and raise both sides to the power 2^{n+1}, then,

$$\frac{\left\{4\sin.\dfrac{\theta}{4^n}\right\}^{2^{n+1}}}{\left\{4\sin.\theta\right\}^2} = \overset{n}{P}\left\{\tan.\frac{\theta}{4^n}\sec.\frac{2\theta}{4^n}\right\}^{\frac{1}{2^n}} =$$

$$= \left\{\tan.\frac{\theta}{4}\sec.\frac{2\theta}{4}\right\}^2.\left\{\tan.\frac{\theta}{16}\sec.\frac{2\theta}{16}\right\}^4 \ldots \left\{\tan.\frac{\theta}{4^n}\sec.\frac{2\theta}{4^n}\right\}^{2^n}\ \ldots (5,2)$$

By assuming $fx = x^2$, $\phi x = \dfrac{1-x^{2a+1}}{1-x} = 1+x+x^2+\&c.\,x^{2a}$,

we should find

$$\psi x = \sqrt[p-1]{1-x+x^2-\&c.+x^{2a}}$$

and by pursuing the same method so frequently repeated, we should arrive at the following results

$$\left\{\frac{1-x^2}{1-x^{2.2a+1}}\left\{\frac{1-x^{2\,(2a+1)}}{1-x^2}\right\}^{\frac{1}{p^n}}\right\}^{\frac{1}{p-1}} = \overset{n}{P}\left\{\frac{1-x^{1.2^n}+x^{2.2^n}-\&c.+x^{2a.2^n}}{\sqrt[p-1]{1+x^{1.2^n}+x^{2.2^n}+\&c.+x^{2a.2^n}}}\right\}^{\frac{1}{p^n}}..(5,3)$$

$$\left\{ \frac{\sin. 2\,\theta}{\sin. 2.\overline{2a+1}\,\theta} \cdot \left(\frac{\sin. 2^{\overline{n+1}}\,\overline{2a+1}\,\theta}{\sin. 2^{n+1}\theta}\right)^{\frac{1}{p^n}} \right\}^{\frac{1}{p-1}} =$$

$$= \overset{n}{P}\left\{\sqrt[p-1]{\frac{\pm\{1-2\cos. 2.2^{n}\,\theta+2\cos. 4.2^{n}\,\theta-\&c.+2\cos. 2a.2^{n}\,\theta\}}{1+2\cos. 2.2^{n}\theta+2\cos. 4.2^{n}\theta+\&c.+2\cos. 2a.2^{n}\theta}}\right\}^{\frac{1}{p^n}} \quad \ldots (5,4)$$

$$\left\{ \frac{\sin. \theta}{\sin. \overline{2a+1}\,\theta} \cdot \left(\frac{\sin. 2^{n}.\overline{2a+1}\,\theta}{\sin. 2^{n}\theta}\right)^{\frac{1}{p^n}} \right\}^{\frac{2}{p-1}} =$$

$$= \overset{n}{P}\left\{\sqrt[p-1]{\frac{(2a+1)-2(2a)\cos. 1.2^{n}\theta+2(2a-1)\cos. 2.2^{n}\theta-\&c.+2(1)\cos. 2a.2^{n}\theta}{(2a+1)+2(2a)\cos. 1.2^{n}\theta+2(2a-1)\cos. 2.2^{n}\theta+\&c.+2(1)\cos. 2a.2^{n}\theta}}\right\}^{\frac{1}{p^n}}(5,5)$$

$\pm$ being used in (5, 4) as a, is an even or an odd number.

In (5, 4) let $a = 1$, $p = 2$,

$$\frac{\sin. 2\,\theta}{\sin. 2.3\theta}\left\{\frac{\sin. 3.2^{n+1}\theta}{\sin. 2^{n+1}\theta}\right\}^{\frac{1}{2^n}} = \overset{n}{P}\left\{\frac{2\cos. 2^{n}.2\theta-1}{2\cos. 2^{n}.2\theta+1}\right\}^{\frac{1}{2^n}} = \overset{n}{P}\left\{\frac{2-\sec. 2^{n}.2\theta}{2+\sec. 2^{n}.2\theta}\right\}^{\frac{1}{2^n}} =$$

$$= \left\{\frac{2-\sec. 4\theta}{2+\sec. 4\theta}\right\}^{\frac{1}{2}} \cdot \left\{\frac{2-\sec. 8\theta}{2+\sec. 8\theta}\right\}^{\frac{1}{4}} \cdots \left\{\frac{2-\sec. 2^{n}.2\theta}{2+\sec. 2^{n}.2\theta}\right\}^{\frac{1}{2^n}} \quad \ldots (5,6)$$

For θ put $\frac{\theta}{2^{n+1}}$, and raise both sides to the 2^{n+1} power,

$$\left\{\frac{\sin. \frac{\theta}{2}}{\sin. \frac{3\theta}{2^{n}}}\right\}^{2^{n+1}} \cdot \left\{\frac{\sin. 3\theta}{\sin. \theta}\right\}^{2} = \overset{n}{P}\left\{\frac{2\cos. \frac{2\theta}{2^{n}}-1}{2\cos. \frac{2\theta}{2^{n}}+1}\right\}^{2} = \overset{n}{P}\left\{\frac{2-\sec. \frac{2\theta}{2^{n}}}{2+\sec. \frac{2\theta}{2^{n}}}\right\}^{\frac{1}{2^n}} =$$

$$= \left\{\frac{2-\sec. \theta}{2+\sec. \theta}\right\}^{2} \cdot \left\{\frac{2-\sec. \frac{\theta}{2}}{2+\sec. \frac{\theta}{2}}\right\}^{2} \cdots \left\{\frac{2-\sec. \frac{\theta}{2^{n-1}}}{2+\sec. \frac{\theta}{2^{n-1}}}\right\}^{2} \quad \ldots (5,7)$$

In (3, 6) make $a = 1$, and also $a = 2$, then,

$$\left\{ \frac{1-x^2}{\left\{1-x^{2^{n+1}}\right\}^{\frac{1}{p^n}}} \right\}^{\frac{1}{p-1}} =$$

E

$$= \left\{\frac{1-x^2}{^{p-1}\sqrt{1+x^2}}\right\}^{\frac{1}{p}} \left\{\frac{1-x^4}{^{p-1}\sqrt{1+x^4}}\right\}^{\frac{1}{p}} \&c. \left\{\frac{1-x^{2^n}}{^{p-1}\sqrt{1+x^{2^n}}}\right\}^{\frac{1}{p}} = \overset{n}{P}\left\{\frac{1-x^{2^n}}{^{p-1}\sqrt{1+x^{2^n}}}\right\}^{\frac{1}{p^n}} \ldots (5,8)$$

$$\sqrt{\frac{1\pm x^3}{\{1\pm x^{3^{n+1}}\}^{\frac{1}{3^n}}}} = \left\{\frac{1\pm x^3}{\sqrt{1\mp x^3 + x^6}}\right\}^{\frac{1}{3}} \left\{\frac{1\pm x^9}{\sqrt{1\mp x^9 + x^{18}}}\right\}^{\frac{1}{3}} \&c. \left\{\frac{1\pm x^{3^n}}{\sqrt{1\mp x^{3^n} + x^{2.3^n}}}\right\}^{\frac{1}{3}} \ldots (5,9)$$

In $(5,3)$ let $p=2$, $a=1$, and multiply the result by $(3,2)$,

$$\frac{x^4+x^2+1}{x^2-1}\left\{\frac{x^{2^{n+1}}-1}{x^{2.2^{n+1}}+x^{1.2^{n+1}}+1}\right\}^{\frac{1}{2^n}} =$$

$$= \left\{\frac{x^6+2x^4+2x^2+1}{x^6-2x^4+2x^2-1}\right\}^{\frac{1}{2}} \left\{\frac{x^{12}+2x^8+2x^4+1}{x^{12}-2x^8+2x^4-1}\right\}^{\frac{1}{2}} \&c. \left\{\frac{x^{3.2^n}+2x^{2.2^n}+2x^{1.2^n}+1}{x^{3.2^n}-2x^{2.2^n}+2x^{1.2^n}+1}\right\}^{\frac{1}{2}} \ldots (6,0)$$

In all the preceding expressions the value of p has not been at all restricted, but may be a whole number or fraction, positive or negative.

They may therefore be considered as variable quantities, and their differential co-efficients taken relative to p.

This will be much facilitated by the aid of the following simple theorem, relating to Continued Products, when differentiated relative to a quantity which only enters as an exponent.

Let v_n and u_n be functions of n and p.

If
$$\overset{n}{P}\left\{a_n{}^{v_n}\right\} = b_n{}^{u_n}$$

Then,
$$\overset{n}{P}\left\{a_n{}^{\frac{d^q.v_n}{(dp)^q}}\right\} = b_n{}^{\frac{d^q.u_n}{(dp)^q}}$$

Which may be easily demonstrated, thus, by taking the logarithm of the first equation,

$$\overset{n}{S}\left\{v_n \log. a_n\right\} = u_n \log. b_n$$

differentiating q times relative to p,

$$\overset{n}{S} \left\{ \frac{d^q . v_n}{(dp)^q} \log. a_n \right\} = \frac{d^q . u_n}{(dp)^q} \log. b_n$$

And reverting from logarithms to numbers,

$$\overset{n}{P} \left\{ a_n^{\frac{d^q . v_n}{(dp)^q}} \right\} = b_n^{\frac{d^q . u}{(dp)^q}} \dots (6,1)$$

If we apply this to $(5,8)$ making $q = 1$, the result will be

$$\frac{\left\{1-x^2\right\}^p}{\left\{1-x^{2^{n+1}}\right\}^{\frac{n.\overline{p-1}+p}{p^n}}} = \overset{n}{P} \left\{ \frac{\left\{1-x^2\right\}^{n.\overline{p-1}}}{\left\{1+x^{2^n}\right\}^{n.\overline{p-1}+p}} \right\}^{\frac{1}{p^n}} \dots (6,2)$$

For x substitute successively xv and xv^{-1}, multiplying the results and putting

$$2 \cos. \theta \quad \text{for} \quad u^{+1} + v^{-1}$$

we have

$$\frac{\left\{1-2x^2 \cos. 2\theta + x^4\right\}^p}{\left\{1-2x^{2^{n+1}} \cos. 2^{n+1}\theta + x^{2^{n+2}}\right\}^{\frac{n.\overline{p-1}+p}{p^n}}} = \overset{n}{P} \left\{ \frac{\left\{1-2x^{2^n} \cos. 2^n\theta + x^{2^{n+1}}\right\}^{n.\overline{p-1}}}{\left\{1+2x^{2^n} \cos. 2^n\theta + x^{2^{n+1}}\right\}^{n.\overline{p-1}+p}} \right\}^{\frac{1}{p^n}} \dots (6,3)$$

Let $x = 1$, for θ put 2θ,

$$\frac{\left\{2 \sin. 2\theta\right\}^{2p}}{\left\{2 \sin. 2^{n+1}\theta\right\}^{\frac{2n.\overline{p-1}+p}{p^n}}} = \overset{n}{P} \left\{ \frac{\left\{2 \sin. 2^n\theta\right\}^{2n.\overline{p-1}}}{\left\{2 \cos. 2^n\theta\right\}^{2(n.\overline{p-1}+p)}} \right\}^{\frac{1}{p^n}} \dots (6,4)$$

Extract the square root, and make $p = 2$,

$$\frac{\left\{2 \sin. 2\theta\right\}^2 \times 4^{1-\frac{1}{2}n}}{\left\{2 \sin. 2^{n+1}\theta\right\}^{\frac{n+2}{2^n}}} = \overset{n}{P} \left\{ (\tan. 2^n\theta)^n (\sec. 2^n\theta)^2 \right\}^{\frac{1}{2}n} =$$

$$= \left\{ \tan. 2\theta)^1 (\sec. 2\theta)^2 \right\}^{\frac{1}{2}} . \left\{ (\tan. 4\theta)^2 (\sec. 4\theta)^2 \right\}^{\frac{1}{4}} \dots \left\{ (\tan. 2^n\theta)^n (\sec. 2^n\theta)^2 \right\}^{\frac{1}{2}n} \dots (6,5)$$

By continuing the differentiations according to formula $(6,1)$, and by making the usual substitutions, we might derive other theorems which would become gradually more complex, when expressed generally, though in particular cases they afford some curious results, these and many of the above might be still further extended by putting $\frac{1}{p}$ for p.

multiplying or dividing the results, and making

$$p^{+1} + p^{-1} = 2 \cos. y$$

this would introduce sines and cosines, &c. into the exponents.

We may obtain some other rather curious Continued Products, by combining the expression

$$\cos. \frac{\theta}{2} = \left\{ \frac{1 + \cos. \theta}{2} \right\}^{\frac{1}{2}}$$

with the expression of Euler's (1, 1)

$$2 \cos. \frac{\theta_{\iota}}{2} = \left\{ 2 + 2 \cos. \theta \right\}^{\frac{1}{2}}$$

$$2 \cos. \frac{\theta}{2^2} = \left\{ 2 + 2 \cos. \frac{\theta}{2} \right\}^{\frac{1}{2}} = \left\{ 2 + \left\{ 2 + 2 \cos. \theta \right\}^{\frac{1}{2}} \right\}^{\frac{1}{2}}$$

$$2 \cos. \frac{\theta}{2^3} = \left\{ 2 + 2 \cos. \frac{\theta}{2^2} \right\}^{\frac{1}{2}} = \left\{ 2 + \left\{ 2 + \left\{ 2 + 2 \cos. \theta \right\}^{\frac{1}{2}} \right\}^{\frac{1}{2}} \right\}^{\frac{1}{2}}$$

and in general

$$2 \cos. \frac{\theta}{2^n} = \left\{ 2 + \left\{ 2 + \left\{ 2 + \&c. \overline{n-1} \text{ terms} \left\{ 2 + 2 \cos. \theta \right\}^{\frac{1}{2}} \right. \right. \right.^{*} \dots (6, 6)$$

Multiply and divide both sides by 2,

$$4 \cos. \frac{\theta}{2^n} = \left\{ 2^{\frac{1}{2}+1} + \left\{ 2^{\frac{2}{2}+1} + \left\{ 2^{\frac{3}{2}+1} + \&c. \left\{ 2^{\frac{n}{2}+1} + 2^{\frac{n}{2}+1} \cos. \theta \right\}^{\frac{1}{2}} \right. \right. \right. \dots (6, 7)$$

$$\cos. \frac{\theta}{2^n} = \left\{ \frac{1}{2^{\frac{1}{2}-1}} + \left\{ \frac{1}{2^{\frac{2}{2}-1}} + \left\{ \frac{1}{2^{\frac{3}{2}-1}} + \&c. \left\{ \frac{1}{2^{\frac{n}{2}-1}} + \frac{1}{2^{\frac{n}{2}-1}} \cos. \theta \right\}^{\frac{1}{2}} \right. \right. \right. \dots (6, 8)$$

---

* When $\theta = o$, the expression $\left\{ 2 + \left\{ 2 + \left\{ 2 + n \text{ terms} \left\{ 2 + 2 \right\}^{\frac{1}{2}} \right.^{\frac{1}{2}} \right.^{\frac{1}{2}} \right.^{\frac{1}{2}}$ is constantly equal to 2, whatever be the value of n. This similar equation also is always true whatever be the value of p, viz.

$$a.\overline{1+a}^{\frac{1}{p-1}} = a \left\{ \overline{a+1}^{\frac{1}{p-1}} + a \left\{ \overline{a+1}^{\frac{1}{p-1}} + a \left\{ \overline{a+1}^{\frac{1}{p-1}} + \&c. \text{ to } n \text{ terms} \left\{ \overline{a+1}^{\frac{1}{p-1}} + a.\overline{a+1}^{\frac{1}{p-1}} \right\} \right.^{\frac{1}{p}} \right.^{\frac{1}{p}} \right.^{\frac{1}{p}}$$

In $(6,6)$ let n vary from 1 to n; then,

$$\overset{n}{P}\left\{2\cos.\frac{\theta}{2^n}\right\} = \overset{n}{P}\left\{\sqrt[7]{2}+\sqrt[7]{2}+\sqrt[7]{2}+\text{ to } n \text{ terms }+2\cos.\theta\right\}$$

but,

$$\overset{n}{P}\left\{2\cos.\frac{\theta}{2^n}\right\} = 2^n \times \overset{n}{P}\left\{\cos.\frac{\theta}{2^n}\right\} = 2^n\,\frac{\sin.\theta}{2^n\sin.\dfrac{\theta}{2^n}} = \frac{\sin.\theta}{\sin.\dfrac{\theta}{2^n}}$$

therefore,

$$\frac{\sin.\theta}{\sin.\dfrac{\theta}{2^n}} = \overset{n}{P}\left\{\sqrt[7]{2}+\sqrt[7]{2}+\sqrt[7]{2}+\text{ to } n \text{ terms }+2\cos.\theta\right\}\ \ldots\ (6,9)$$

and similarly, from $(6,7)$ and $(6,8)$, we shall have

$$\frac{2^n\sin.\theta}{\sin.\dfrac{\theta}{2^n}} = \overset{n}{P}\left\{\sqrt[7]{2^{2}+1}+\sqrt[7]{2^{2}+1}+\sqrt[7]{2^{2}+1}+\&\text{c.}+\sqrt[7]{2^{2}+1}+2^{2}+1\cos.\theta\right\}\ \ldots\ (7,0)$$

and,

$$\frac{\sin.\theta}{2^n\sin.\dfrac{\theta}{2^n}} = \overset{n}{P}\left\{\sqrt{\frac{1}{2^{2}-1}}+\sqrt[4]{\frac{1}{2^{2}+1}}+\sqrt[4]{\frac{1}{2^{2}+1}}+\&\text{c.}\sqrt{\frac{1}{2^{2}-1}}+\frac{1}{2^{2}-1}\cos.\theta\right\}\ \ldots\ (7,1)$$

Many of the above theorems have a finite value, when the product is continued to infinity; thus, from (2) (5),

$$\frac{1}{1-x^2} = \overset{n}{P}\left\{1+x^{2^n}\right\} \qquad \begin{bmatrix}n=1\\n=\infty\end{bmatrix}\ldots.(7,2)$$

$$\frac{1}{1-x^3} = \overset{n}{P}\left\{1-x^{3^n}+x^{2.3^n}\right\} \qquad \begin{bmatrix}n=1\\n=\infty\end{bmatrix}\ldots.(7,3)$$

If we meet with a quantity of this kind, $\dfrac{\sin.\dfrac{a\theta}{c^n}}{\sin.\dfrac{b\theta}{c^n}}$ its value, when n becomes infinite, may be easily found; thus,

$$\frac{\sin.\dfrac{a\theta}{c^n}}{\sin.\dfrac{b\theta}{c^n}} = \frac{\dfrac{a\theta}{c^n}\cdot\dfrac{1}{1}-\dfrac{(a\theta)^3}{c^{3n}}\cdot\dfrac{1}{1.2.3}+\&\text{c.}}{\dfrac{b\theta}{c^n}\cdot\dfrac{1}{1}-\dfrac{(b\theta)^3}{c^{3n}}\cdot\dfrac{1}{1.2.3}+\&\text{c.}} = \frac{a\theta-\dfrac{(\dot a\theta)^3}{c^{2n}}\cdot\dfrac{1}{1.2.3}+\&\text{c.}}{b\theta-\dfrac{(b\theta)^3}{c^{2n}}\cdot\dfrac{1}{1.2.3}+\&\text{c.}}$$

F

and when n becomes infinite, this becomes

$$\frac{\sin. \dfrac{a\theta}{c^n}}{\sin. \dfrac{b\theta}{c^n}} = \frac{a}{b}$$

therefore $(2,3)$ and $(2,4)$ become

$$\cos.\theta = \overset{n}{P} \left\{ 1 - 2\cos.\frac{2\theta}{(4a\pm1)^n} + 2\cos.\frac{4\theta}{(4a\pm1)^n} - \&c. + 2\cos.\frac{4a\theta}{(4a+1)^n} \right\} \begin{bmatrix} n = 1 \\ n = \infty \end{bmatrix} \cdot(7,4)$$

$$\cos.\theta = \overset{n}{P} \left\{ 2\cos.\frac{\overline{4a-2}\,\theta}{(4a-1)^n} - 2\cos.\frac{\overline{4a-4}\,\theta}{(4a-1)^n} + \&c. + 2\cos.\frac{2\theta}{(4a-1)^n} - 1 \right\} \begin{bmatrix} n = 1 \\ n = \infty \end{bmatrix} \cdot(7,5)$$

$(2,5)$ and $(2,6)$ are, when n is infinite,

$$\frac{\sin. \overline{4a+1}\,\theta}{\overline{4a+1}\,\sin.\theta} = \overset{n}{P} \left\{ 1 - 2\cos.\frac{2\theta}{2^n} + 2\cos.\frac{4\theta}{2^n} - \&c. + 2\cos.\frac{4a\theta}{2^n} \right\} \begin{bmatrix} n = 1 \\ n = \infty \end{bmatrix} \cdot\cdot(7,6)$$

$$\frac{\sin. \overline{4a-1}\,\theta}{\overline{4a-1}\,\sin.\theta} = \overset{n}{P} \left\{ 2\cos.\frac{\overline{4a-2}\,\theta}{2^n} - 2\cos.\frac{\overline{4a-4}\,\theta}{2^n} + \&c. + 2\cos.\frac{2\theta}{2^n} - 1 \right\} \begin{bmatrix} n = 1 \\ n = \infty \end{bmatrix} \cdot(7,7)$$

All the preceding Continued Products have been accurately determined for any given value of (n), those which follow are only determined when n is infinite, and as these depend on a principle entirely different from the others, I have chosen to place them last.

Let
$$\phi x = \underset{0}{A} + \underset{1}{A}x + \underset{2}{A}x^2 + \&c. \underset{n}{A}x^n$$

For x, substitute successively x, x^2, x^3, &c, ad infinitum;

$$\phi x = \underset{0}{A} + \underset{1}{A}x + \underset{2}{A}x^2 + \&c. \underset{n}{A}x^n$$
$$\phi x^2 = \underset{0}{A} + \underset{1}{A}x^2 + \underset{2}{A}x^4 + \&c. \underset{n}{A}x^{2n}$$
$$\phi x^3 = \underset{0}{A} + \underset{1}{A}x^3 + \underset{2}{A}x^6 + \&c. \underset{n}{A}x^{3n}$$
$$\phi x^4 = \underset{0}{A} + \underset{1}{A}x^n + \underset{2}{A}x^8 + \&c. \underset{n}{A}x^{in}$$
$$\&c. \qquad \&c. \qquad \&c.$$

Add all the vertical terms, and suppose every even line negative; then,

$$(\phi x - \underset{0}{A}) - (\phi x^2 - \underset{0}{A}) + (\phi x^3 - \underset{0}{A}) - \&c. \text{ ad inf. } =$$

$$= \frac{\underset{1}{A}x}{1+x} + \frac{\underset{2}{A}x^2}{1+x^2} + \frac{\underset{3}{A}x^3}{1+x^3} + \&c. \frac{\underset{n}{A}x^n}{1+x^n} \ldots (a,1)$$

Supposing all positive,

$$(\phi x - \underset{0}{A}) + (\phi x^2 - \underset{0}{A}) + (\phi x^3 - \underset{0}{A}) + \&c.\ \text{ad inf.} =$$

$$= \frac{\underset{1}{A}\, x}{1-x} + \frac{\underset{2}{A}\, x^2}{1-x^2} + \frac{\underset{3}{A}\, x^3}{1-x^3} + \&c.\ \frac{\underset{n}{A}\, x^n}{1-x^n} \ldots (a, 2)$$

Multiply $(a, 1)$ $(a, 2)$ by $\dfrac{dx}{x}$, and let

$$\int (\phi x - \underset{0}{A})\, \frac{dx}{x} = fx$$

then,

$$\int (\phi x^n - \underset{0}{A})\, \frac{dx}{x} = \frac{1}{n} \int (\phi x^n - \underset{0}{A})\, \frac{n x^{n-1} dx}{x^n} = \frac{1}{n} f x^n$$

therefore,

$$\frac{fx}{1} - \frac{fx^2}{2} + \frac{fx^3}{3} - \frac{fx^4}{4} + \&c. = S\left\{ -(-1)^n \frac{fx^n}{n} \right\} =$$

$$= \log.\left\{ (1+x)^{\frac{1}{1}\underset{1}{A}} \cdot (1+x^2)^{\frac{1}{2}\underset{2}{A}} \cdot (1+x^3)^{\frac{1}{3}\underset{3}{A}} \cdot\, \&c.\ (1+x^n)^{\frac{1}{n}\underset{n}{A}} \right\} \ldots (a, 3)$$

$$\frac{fx}{1} + \frac{fx^2}{2} + \frac{fx^3}{3} + \&c.\ \text{ad inf.} = S\left(\frac{fx^n}{n} \right)$$

$$- \log.\left\{ (1-x)^{\frac{1}{1}\underset{1}{A}} \cdot (1-x^2)^{\frac{1}{2}\underset{2}{A}} \cdot (1-x^3)^{\frac{1}{3}\underset{3}{A}} \cdot \ldots (1-x^n)^{\frac{1}{n}\underset{n}{A}} \right\} \ldots (a, 4)$$

The sum of these two is

$$\frac{fx}{1} + \frac{fx^3}{3} + \frac{fx^5}{5} + \&c.\ \text{ad inf.}$$

$$- \frac{1}{2} \log.\left\{ \left(\frac{1-x}{1+x}\right)^{\frac{1}{1}\underset{1}{A}} \cdot \left(\frac{1-x^2}{1+x^2}\right)^{\frac{1}{2}\underset{2}{A}} \cdot \left(\frac{1-x^3}{1+x^3}\right)^{\frac{1}{3}\underset{3}{A}} \cdots \left(\frac{1-x^n}{1+x^n}\right)^{\frac{1}{n}\underset{n}{A}} \right\} \ldots (a, 5)$$

These expressions, though they apparently give the value to n terms, are in fact of very little use, except when n is infinite. In that case, the value of the series on the left hand may frequently be obtained.

In $(a, 4)$ make

$$\underset{0}{A} = 0 \quad \underset{1}{A} = 1 \quad \underset{2}{A} = 2 \quad \underset{3}{A} = 3, \text{ &c.}$$

$$\phi x = \frac{x}{(1-x)^2}$$

$$fx = \int (\phi x - \underset{0}{A}) \frac{dx}{x} = \int \frac{dx}{(1-x)^2} = \frac{1}{(1-x}$$

$$S \left\{ \frac{f x^n}{n} \right\} = + \left\{ \frac{1}{1-x} \cdot \frac{1}{1} + \frac{1}{1-x^2} \cdot \frac{1}{2} + \frac{1}{1-x^3} \cdot \frac{1}{3} + \text{&c.} \right\} + C =$$

$$= - \log. \left\{ (1-x) \cdot (1-x^2) \cdot (1-x^3) \ldots \text{&c.} \right\}$$

If $x = 0$,

$$- C = \frac{1}{1} + \frac{1}{2} + \frac{1}{3} + \text{&c.}$$

$$- \left\{ \frac{1}{1} \left(1 - \frac{1}{1-x} \right) + \frac{1}{2} \left(1 - \frac{1}{1-x^2} \right) + \frac{1}{3} \left(1 - \frac{1}{1-x^3} \right) + \text{&c.} \right\} =$$

$$= + \frac{x}{1-x} \cdot \frac{1}{1} + \frac{x^2}{1-x^2} \cdot \frac{1}{2} + \frac{x^3}{1-x^3} \cdot \frac{1}{3} + \text{&c.} =$$

$$- \log. \left\{ (1-x) \cdot (1-x^2) \cdot (1-x^3) \ldots \text{ad inf.} \right\}$$

$$- \left\{ \frac{x}{1} \cdot \frac{1}{1} + \frac{x^2}{1+x} \cdot \frac{1}{2} + \frac{x^3}{1+x+x^2} \cdot \frac{1}{3} + \text{&c.} \right\} =$$

$$= \log. \left\{ (1-x)(1-x^2)(1-x^3) \ldots \text{ad inf.} \right\}^{1-x} \ldots (a, 6$$

If x ultimately equal unity,

$$\varepsilon^{-\left\{ \frac{1}{1^2} + \frac{1}{2^2} + \frac{1}{3^2} + \text{&c.} \right\}} = \left\{ (1-x)(1-x^2)(1-x^3) \ldots \text{ad inf.} \right\}^{1-x} \ldots (a, 7)$$

$$\varepsilon^{-\frac{\pi^2}{6}} = \left\{ (1-x)(1-x^2)(1-x^3) \ldots \text{ad inf.} \right\}^{1-x} \qquad x \text{ ult.} = 1$$

$$\varepsilon^{-\frac{\pi^2}{6}} = \left\{ (1)(1+x)(1+x+x^2) \ldots \text{ad inf.} \right\}^{1-x} \times \log. \left\{ (1-x)^{\frac{1}{1-x}} \right\}^{1-x}$$

$$\frac{\varepsilon^{-\frac{\pi^2}{6}}}{1-x} = \left\{1 \cdot 2 \cdot 3 \cdot 4 \dots \text{ad inf.}\right\}^{(1-x)} x \text{ ult.} = 1$$

In $(a, 4)$, let $\underset{0}{A} = 0$, $\underset{1}{A} = n'$, $\underset{2}{A} = n''$, $\underset{3}{A} = n'''$, &c.

n' n'' n''', &c. being the successive terms of the nth order of figurate numbers.

$$\phi x = \frac{x}{(1-x)^n} \quad f x = \frac{1}{(n-1)(1-x)^{n-1}}$$

$$\frac{1}{m} f x^m = \frac{\overline{n-1}^{-1}}{m \cdot (1-x)^{n-1}}$$

$$\frac{1}{n-1} \left\{ \frac{1^{-1}}{(1-x)^{n-1}} + \frac{2^{-1}}{(1-x^2)^{n-1}} + \frac{3^{-1}}{(1-x^3)^{n-1}} + \text{&c.} \right\} + C =$$

$$= -\log. \left\{ (1-x)^{\frac{n'}{1}} \cdot (1-x^2)^{\frac{n''}{2}} \cdot (1-x^3)^{\frac{n'''}{3}} \dots \text{ad inf.} \right\}$$

if $x = 0$, $C = -\dfrac{1}{n-1} \left\{ \dfrac{1}{1} + \dfrac{1}{2} + \dfrac{1}{3} + \text{&c.} \right\}$

$$-\frac{1}{n-1} \left\{ \frac{1-(1-x)^{n-1}}{1 \cdot (1)^{n-1}} + \frac{1-(1-x^2)^{n-1}}{2 \cdot (1+x)^{n-1}} + \frac{1-(1-x^3)^{n-1}}{3 \cdot (1+x+x^2)^{n-1}} + \text{&c.} \right\}$$

$$\varepsilon \dots \dots \dots \dots \dots \dots \dots \dots \dots \dots \dots \dots \dots \dots \dots \dots \dots \dots =$$

$$= \left\{ (1-x)^{\frac{n'}{1}} \cdot (1-x^2)^{\frac{n''}{2}} \cdot (1-x^3)^{\frac{n'''}{3}} \dots \text{ad inf.} \right\}^{(1-x)^{n-1}}$$

Let x ultimately become unity,

$$\varepsilon^{-\frac{1}{n-1} \left\{ \frac{1}{1^n} + \frac{1}{2^n} + \frac{1}{3^n} + \text{&c.} \right\}} \dots \dots \dots \dots = \left\{ (1-x)^{\frac{n'}{1}} \cdot (1-x^2)^{\frac{n''}{2}} \cdot (1-x^3)^{\frac{n'''}{3}} \dots \text{ad inf.} \right\}^{(1-x)^{n-1}} \dots \dots (a, 8)$$

or, $\dfrac{\varepsilon^{-\frac{1}{n-1} \left\{ \frac{1}{1^n} + \frac{1}{2^n} + \frac{1}{3^n} + \text{&c.} \right\}}}{\left\{ 1-x \right\}^{\frac{1}{n-1}}} \dots \dots \dots \dots \dots = \left\{ (1)^{\frac{n'}{1}} \cdot (2)^{\frac{n''}{2}} \cdot (3)^{\frac{n'''}{3}} \dots \text{ad inf.} \right\}^{(1-x)^{n-1}} \dots \dots (a, 9)$

When n is an even number,

$$\varepsilon^{-\frac{1}{n-1}\cdot\frac{2^{n-1}B_{n-1}\pi^n}{1.2.3.4\ldots n}}\ldots\ldots\ldots = \left\{(1-x)^{\frac{n'}{1}}.\ (1-x^2)^{\frac{n''}{2}}.\ (1-x^3)^{\frac{n'''}{3}}.\ \&\text{c.}\right\}^{(1-x)^{n-1}}$$

B_n being one of the numbers of Bernouilli.

Let, $\quad \underset{0}{A} = 0,\ \underset{1}{A} = 1,\ \underset{2}{A} = 0,\ \underset{3}{A} = -1,\ \underset{4}{A} = 0,\ \underset{5}{A} = 1,\ \&\text{c.}$

$$\phi x = x - x^3 + x^5 - \&\text{c.} = \frac{x}{1+x^2}$$

$$fx = \text{arc (tan.} = x) = \tan.^{-1}x \ *$$

Substituting these values in $(a, 3)$ $(a, 4)$, we shall find

$$\frac{\tan.^{-1}x}{1} + \frac{\tan.^{-1}x^2}{2} + \frac{\tan.^{-1}x^3}{3} + \&\text{c.} =$$

$$= -\log.\ \overset{n}{P}\left\{(1 - x^{\overline{2n-1}})^{\frac{-(-1)^n}{2n-1}}\right\} = -\log.\left\{(1-x)^{\frac{1}{1}}.(1-x^3)^{-\frac{1}{3}}.(1-x^5)^{\frac{1}{5}}..\text{ad inf.}\right\}.\ (a, 10)$$

$$\frac{\tan.^{-1}x}{1} - \frac{\tan.^{-1}x^2}{2} + \frac{\tan.^{-1}x^3}{3} - \&\text{c.} =$$

$$= \log.\ \overset{n}{P}\left\{(1 + x^{\overline{2n-1}})^{\frac{-(-1)^n}{2n-1}}\right\} = \log.\left\{(1+x)^{\frac{1}{1}}.(1+x^3)^{-\frac{1}{3}}.(1+x^5)^{\frac{1}{5}}..\text{ad inf.}\right\}\ldots.\ (a, 11)$$

In $(a, 10)$, let $x = 1$,

$$\frac{\pi}{4}\left(\frac{1}{1} + \frac{1}{2} + \frac{1}{3} + \&\text{c.}\right) =$$

$$= -\log.\ (1-x)^{\frac{\pi}{4}} - \log.\left\{(1)^{\frac{1}{1}}.\ (1+x+x^2)^{-\frac{1}{3}}.\ (1+x+x^2+x^3+x^4)^{\frac{1}{5}}.\ \text{ad inf.}\right\}$$

$$-\frac{\pi}{4}\log.\ (\overline{1-x}) + \frac{\pi}{4}\log.\ (1-x) = -\log.\left\{(1)^{\frac{1}{1}}.(3)^{-\frac{1}{3}}.(5)^{\frac{1}{5}}..\text{ad inf.}\right\}$$

$$1 = \left\{(1)^{\frac{1}{1}}.\ (3)^{-\frac{1}{3}}.\ (5)^{\frac{1}{5}}.\ (7)^{-\frac{1}{7}}\ldots\text{ad inf.}\right\}\ldots.\ (a, 12)$$

---

* $\tan.^{-1}x$ or, $\sin.^{-1}x$, signifies the arc whose tangent or sine is equal to x.

Let $\phi x = \dfrac{x}{\sqrt{1-x^2}}$,

$$fx = \sin.^{-1}x$$

$$\underset{0}{A} = 0, \ \underset{1}{A} = 1, \ \underset{2}{A} = \frac{1}{2}, \ \underset{3}{A} = \frac{1.3}{2.4} \ . \ \&c.$$

$$\frac{\sin.^{-1}x}{1} + \frac{\sin.^{-1}x^2}{2} + \frac{\sin.^{-1}x^3}{3} + \&c. =$$

$$= - \log. \left\{ (1-x)^{1\cdot\frac{1}{1}}.(1-x^2)^{\frac{1}{2}\cdot\frac{1}{2}}.(1-x^3)^{\frac{1\cdot3}{2\cdot4}\cdot\frac{1}{3}}. \text{ ad inf.} \right\} \ldots (a, 13)$$

$$\frac{\sin.^{-1}x}{1} - \frac{\sin.^{-1}x^2}{2} + \frac{\sin.^{-1}x^3}{3} - \&c. =$$

$$= \log. \left\{ (1+x)^{1\cdot\frac{1}{1}}.(1+x^2)^{\frac{1}{2}\cdot\frac{1}{2}}.(1+x^3)^{\frac{1\cdot3}{2\cdot4}\cdot\frac{1}{3}}. \text{ ad inf.} \right\} \ldots (a, 14)$$

A number of particular cases might be produced affording some remarkable results, but we will rather proceed at once to some theorems yet more general.

Assume

$$\phi(x, \theta) = \underset{1}{A} x \cos. \theta + \underset{2}{A} x^2 \cos. 2\theta + \&c. = \overset{n}{S} \left\{ \underset{n}{A} x^n \cos. n\theta \right\}$$

the limits of n being unity and infinite.

Multiply this by $\dfrac{u}{1}$; for θ put 2θ, and multiply by $\dfrac{u^2}{2}$; again, for θ put 3θ, and multiply by $\dfrac{u^3}{3}$, and so on; then we find,

$$\frac{u}{1} \phi(x, \theta) = \frac{u}{1} \overset{n}{S} \left\{ \underset{n}{A} x^n \cos. n\theta \right\}$$

$$\frac{u^2}{2} \phi(x, 2\theta) = \frac{u^2}{2} \overset{n}{S} \left\{ \underset{n}{A} x^n \cos. 2n\theta \right\}$$

$$\frac{u^3}{3} \phi(x, 3\theta) = \frac{u^3}{3} \overset{n}{S} \left\{ \underset{n}{A} x^n \cos. 3n\theta \right\}$$

$$\&c. \qquad\qquad \&c.$$

Adding these together,

$$\overset{p}{S}\left\{\frac{u^p}{p}\,\phi\,(x,p\theta)\right\}=\overset{n}{S}\left\{\underset{n}{A}x^n\times\overset{p}{S}\left\{\frac{u^p}{p}\,\cos.\,pn\theta\right\}\right\}$$

But,

$$\overset{p}{S}\left\{\frac{u^p}{p}\,\cos.\,pn\theta\right\}=\frac{u}{1}\cos.\,n\theta+\frac{u^2}{2}\cos.\,2n\theta+\frac{u^3}{3}\cos.\,3n\theta+\&c.$$

$$=-\frac{1}{2}\log.\left\{1-2u\cos.\,n\theta+u^2\right\}$$

$$\overset{p}{S}\left\{\frac{u^p}{p}\,\phi\,(x,p\theta)\right\}=-\frac{1}{2}\overset{n}{S}\left\{\underset{n}{A}\,x^n\log.\left\{1-2u\cos.\,n\theta+u^2\right\}\right\}=$$

$$-\frac{1}{2}\overset{n}{S}\log.\left\{\left\{1-2u\cos.\,n\theta+u^2\right\}^{\underset{n}{A}x^n}\right\}=$$

$$=-\frac{1}{2}\log.\overset{n}{P}\left\{\left\{1-2u\cos.\,n\theta+u^2\right\}^{n^{x^n}}\right\}$$

Making u negative, and dividing by the result, we have the three expressions,

$$\varepsilon^{-2\overset{p}{S}\left\{\frac{u^p}{p}\,\phi\,(x,\,p\,\theta)\right\}}=\overset{n}{P}\left\{\left\{1-2u\cos.\,n\theta+u^2\right\}^{\underset{n}{A}x^n}\right\}\begin{bmatrix}p=1\\p=\infty\\n=1\\n=\infty\end{bmatrix}\,\ldots\,(a,15)$$

$$\varepsilon^{2\overset{p}{S}\left\{\frac{-(-u)^p}{p}\,\phi\,(x,\,p\,\theta)\right\}}=\overset{n}{P}\left\{\left\{1+2u\cos.\,n\theta+u^2\right\}^{\underset{n}{A}x^n}\right\}\begin{bmatrix}p=1\\p=\infty\\n=1\\n=\infty\end{bmatrix}\,\ldots\,(a,16)$$

$$\varepsilon^{-4\overset{p}{S}\left\{\frac{u^{2p-1}}{2p-1}\,\phi\,(x,\,2p-1\,\theta)\right\}}=\overset{n}{P}\left\{\left\{\frac{1-2u\cos.\,n\theta+u^2}{1+2u\cos.\,n\theta+u^2}\right\}^{\underset{n}{A}x^n}\right\}\begin{bmatrix}p=1\\p=\infty\\n=1\\n=\infty\end{bmatrix}\,\ldots\,(a,17)$$

If $u = 1$, since p and n are between the same limits, we have, by extracting the square root,

$$\varepsilon^{-\overset{n}{S}\left\{\frac{\phi\,(x,\,n\,\theta)}{n}\right\}} \cdots\cdots = \overset{n}{P}\left\{2\sin.\,n\,\frac{\theta}{2}\right\}^{\overset{A}{n}x^n} \quad \begin{bmatrix} n = 1 \\ n = \infty \end{bmatrix} \cdots\cdots (a,\,18)$$

$$\varepsilon^{\,S\left\{\frac{-\,(-1)^n\,\phi\,(x,\,n\,\theta)}{n}\right\}} \cdots\cdots\cdots = \overset{n}{P}\left\{2\cos.\,n\,\frac{\theta}{2}\right\}^{\overset{A}{n}x^n} \quad \begin{bmatrix} n = 1 \\ n = \infty \end{bmatrix} \cdots\cdots (a,\,19)$$

$$\varepsilon^{-\frac{1}{2}\,S\left\{\frac{\phi\,(x,\,\overline{2n-1}\,\theta)}{2n-1}\right\}} \cdots\cdots\cdots = \overset{n}{P}\left\{\tan.\,n\,\frac{\theta}{2}\right\}^{\overset{A}{n}x^n} \quad \begin{bmatrix} n = 1 \\ n = \infty \end{bmatrix} \cdots\cdots (a,\,20)$$

In $(a, 16)$ let $x = 1$, $\underset{n}{A} = 1$,

$$\phi\,(x,\,\theta) = \cos.\,\theta + \cos.\,2\theta + \cos.\,3\theta + \&\text{c.} = -\frac{1}{2}$$

for θ put 2θ,

$$\overset{p}{S}\left\{\frac{-\,(-u)^p}{p}\,\phi\,(x,\,\theta)\right\} = -\frac{1}{2}\log.\,\overline{1+u}$$

$$\varepsilon^{-\log.\,\overline{1+u}} = \frac{1}{1+u} = \overset{n}{P}\left\{1 + 2u\cos.\,2\,n\,\theta + u^2\right\}$$

If $u = 1$,

$$\frac{1}{\sqrt{2}} = \overset{n}{P}\left\{2\cos.\,n\theta\right\} = (2\cos.\,\theta).(2\cos.\,2\,\theta).(2\cos.\,3\,\theta)\ldots\text{ad inf.}$$

Let
$$\underset{1}{A} = \frac{1}{1},\quad \underset{2}{A} = \frac{1}{1.2},\quad \underset{3}{A} = \frac{1}{1.2.3},\quad \&\text{c.}$$

$$\phi\,x = \frac{x\cos.\,\theta}{1} + \frac{x^2\cos.\,2\,\theta}{1.2} + \frac{x^3\cos.\,3\,\theta}{1.2.3} + \&\text{c.} = \varepsilon^{\,x\cos.\,\theta}\cos.\,(x\sin.\,\theta) - 1$$

Taking the logarithms of $(a, 15)$ $(a, 16)$ $(a, 17)$,

$$-2\log.\,\overline{1-u} - 2\left\{\frac{\varepsilon^{\,x\cos.\,\theta}\cos.\,(x\sin.\,\theta)}{1}\,u + \frac{\varepsilon^{\,x\cos.\,2\theta}\cos.\,(x\sin.\,\theta)}{2}\,u^2 + \frac{\varepsilon^{\,x\cos.\,3\theta}\cos.\,(x\sin.\,3\theta)}{3}\,u^3 + \&\text{c.}\right\} =$$

H

$$= \log. \overset{n}{P} \left\{1 - 2u\cos. n\theta + u^2\right\}^{\frac{x^n}{1.2.3...n}} =$$

$$= \log. \left\{\overline{1 - 2u\cos.\theta + u^2}^{\frac{x}{1}} \overline{1 - 2u\cos. 2\theta + u^2}^{\frac{x}{2}} \overline{1 - 2u\cos. 3\theta + u^2}^{\frac{x}{3}} \left\{\text{ad inf.}\right\}\right.$$

$$-2\log.\overline{1+u} + 2\left\{\frac{\varepsilon^{x\cos.\theta}\cos.(x\sin.\theta)}{1} u - \frac{\varepsilon^{x\cos. 2\theta}\cos.(x\sin. 2\theta)}{2} u^2 + \frac{\varepsilon^{x\cos. 3\theta}\cos.(x\sin.\div 3\theta)}{3} u^3 - \&c.\right\}$$

$$= \log. \overset{n}{P} \left\{1 + 2u\cos. n\theta + u^2\right\}^{\frac{x^n}{1.2.3...n}} =$$

$$= \log. \left\{1 + 2u\cos.\theta + u^2\right\}^{\frac{x}{1}} \left\{1 + 2u\cos. 2\theta + u^2\right\}^{\frac{x}{2}} \left\{1 + 2u\cos. 3\theta + u^2\right\}^{\frac{x}{3}} \text{ad inf.}$$

Also,

$$-4\left\{\frac{\varepsilon^{x\cos.\theta}\cos.(x\sin.\theta)}{1} u + \frac{\varepsilon^{x\cos. 3\theta}\cos.(x\sin. 3\theta)}{3} u^3 + \frac{\varepsilon^{x\cos. 5\theta}\cos.(x\sin. 5\theta)}{5} + \&c.\right\} =$$

$$= \log. \overset{n}{P} \left\{\frac{1 - 2u\cos. n\theta + u^2}{1 + 2u\cos. n\theta + u^2}\right\}^{\frac{x^n}{1.2.3...n}} =$$

$$= \log. \left\{\frac{1 - 2u\cos.\theta + u^2}{1 + 2u\cos.\theta + u^2}\right\}^{\frac{x}{1}} \left\{\frac{1 - 2u\cos. 2\theta + u^2}{1 + 2u\cos. 2\theta + u^2}\right\}^{\frac{x}{2}} \left\{\frac{1 - 2u\cos. 3\theta + u^2}{1 + 2u\cos. 3\theta + u^2}\right\}^{\frac{x}{3}} \left\{\&c.\text{ ad inf.}\right\}$$

In. the second of these expressions if u be made equal to unity, and also x and 2θ be put for θ, we have,

$$-\log. 2 + \frac{\varepsilon^{\cos. 2\theta}\cos.\sin. 2\theta}{1} + \frac{\varepsilon^{\cos. 4\theta}\cos.\sin. 4\theta}{2} + \frac{\varepsilon^{\cos. 6\theta}\cos.\sin. 6\theta}{3} - \&c. =$$

$$= \log. \left\{2\cos.\theta\right\}^{\frac{1}{1}} \left\{2\cos. 2\theta\right\}^{\frac{1}{2}} \left\{2\cos. 3\theta\right\}^{\frac{1}{3}} \left\{2\cos. 4\theta\right\}^{\frac{1}{4}} \&c.\text{ ad inf.} \ldots (a, 21)$$

Several of the expressions deduced in the latter part of this paper, are purposely left in their most general form ; should any one enquire into the results produced, by assigning particular values to some of the quantities concerned, he will meet with many singular and curious theorems.

The formulæ in the former part, may by differentiating and by other methods be converted into infinite series, and from these and the expressions themselves, may be derived several remarkable properties of the circle.

TRIGONOMETRICAL SERIES;

Particularly those whose Terms are multiplied by the Tangents, Co-tangents, Secants, &c. of Quantities in Arithmetic Progression: together with some singular Transformations.

———

Tʜᴇ application of Moivre's Theorems enables us with little difficulty to multiply the successive terms of any series whose sum is known, by the sines or cosines of the terms of any proposed arithmetic progressions. We are thus conducted to the general formulæ:

$$\frac{f\{x \cdot \varepsilon^{\theta \sqrt{(-1)}}\} + f\{x \cdot \varepsilon^{-\theta\sqrt{(-1)}}\}}{2} - A_0 = A_1 x \cdot \cos. \theta + A_2 x^2 \cdot \cos. 2\theta + \&c.;$$

and,

$$\frac{f\{x \cdot \varepsilon^{\theta\sqrt{(-1)}}\} - f\{x \cdot \varepsilon^{-\theta\sqrt{(-1)}}\}}{2\sqrt{(-1)}} = A_1 x \cdot \sin. \theta + A_2 x^2 \cdot \sin. 2\theta + \&c.;$$

where $f\{x\}$ is any function of x developable in the form

$$A_0 + A_1 x + A_2 x^2 + \&c.;$$

and,

$$\varepsilon = 1 + \frac{1}{1} + \frac{1}{1 \cdot 2} + \frac{1}{1 \cdot 2 \cdot 3} + \&c.$$

By the help of these equations, and the resolution of $(\cos. i\theta)^n$ and $(\sin. i\theta)^n$ into the cosines or sines of the multiples of $i\theta$, we may with the same facility express the sums of the series,

$$A_0 + A_1 x \cdot (\cos. \theta)^n + A_2 x^2 \cdot (\cos. 2\theta)^n + \&c.$$

$$0 + A_1 x \cdot (\sin. \theta)^n + A_2 x^2 \cdot (\sin. 2\theta)^n + \&c.$$

When however we come to the reverse operation, that of dividing the successive terms by such sines, &c., the case is entirely altered, and no general method has yet been discovered by which this may be accomplished. To this may be referred the general summation of such series as

I

$$\underset{0}{A} + \underset{1}{A}x \cdot \tan \theta + \underset{2}{A}x^2 \cdot \tan 2\theta + \&c.$$
$$\underset{0}{A} + \underset{1}{A}x \cdot \sec \theta + \underset{2}{A}x^2 \cdot \sec 2\theta + \&c.$$
$$\&c. \qquad\qquad \&c. \qquad\qquad \&c.$$

Very little is in fact known of such series;—their discussion is at once laborious and ungrateful, and all that has been hitherto done, appears only to have placed in a clearer light the extreme difficulty of the subject.

In the present Memoir it is proposed to exhibit a sufficiently simple method of summing many series of these and other forms, for particular values of θ. The results will serve to give some idea of the complication attendant upon generality in this point.

But before we proceed to this subject, it will be necessary to say a few words on the Notation employed; a considerable departure from that in common use not being allowable on arbitrary, or capricious grounds.

Let $f(x)$ represent the result of an operation performed on x; $f^2(x)$ may be elegantly used to denote $f(f(x))$, or a repetition of the same operation, and generally

$$f^{m+n}(x) = f^m f^n(x).$$

If now we make $n = -m$, we have

$$f^m f^{-m} x = f^o x = x \ldots\ldots\ldots \text{(See Note 1.)}$$

Hence, $f^{-m}x$ must be such a quantity, that its mth function (f) shall be x; that is, the negative index denotes the reverse operation. Now, in the course of analytical investigations, it is indispensably necessary to consider $\sin x$, $\cos x$, $\log x$, &c., as mere numbers, functions of the number x, and by following up the foregoing idea, we shall have

$$\sin.^2 x = \sin. \sin. x; \ldots\ldots \sin.^{-1}x = \text{number whose sine is } x$$
$$\tan.^{-1}x = \text{number whose tan. is } x$$
$$\log.^{-1}x = \text{number whose log. is } x, \text{ or, } \log.^{-1}x = \varepsilon^x$$

In like manner *,

$$\log.^{-2}x = \varepsilon^{\varepsilon^x}; \ldots. \log.^{-n}x = \varepsilon^{\varepsilon^{\cdot^{\cdot^{\varepsilon^{\varepsilon^x}}}}}_{(n)}; \ldots\ldots \varepsilon^{a\varepsilon^x} = \{\log.^{-2}x\}^a; \ldots. \&c.$$

---

* This mode of writing the inverse logarithms, will be found extremely convenient in expressions much complicated with indices.

It is not however enough to have explained a novel algorithm; the Reader must become familiar with it, and this before he begins to employ it in actual investigation. For this purpose, and to avoid the necessity of making any farther mention of the subject, we shall premise a few simple expressions of great utility:

$$1. -\ \ldots\ldots \tan.^{-1} A = \frac{1}{2\sqrt{(-1)}} \cdot \log. \left\{ \frac{1+A\sqrt{(-1)}}{1-A\sqrt{(-1)}} \right\}$$

$$2. -\ \ldots\ldots \tan.^{-1} A + \tan.^{-1} B = \tan.^{-1} \left\{ \frac{A+B}{1-AB} \right\}$$

$$3. -\ \ldots\ldots \frac{\tan.^{-1}\{x.\varepsilon^{\theta\sqrt{(-1)}}\} - \tan.^{-1}\{x.\varepsilon^{-\theta\sqrt{(-1)}}\}}{\sqrt{(-1)}} = \frac{1}{2}\log. \left\{ \frac{1+2x.\sin.\theta+x^2}{1-2x.\sin.\theta+x^2} \right\},$$

and when $x = 1$,

$$\frac{\tan.^{-1}\{\varepsilon^{\theta\sqrt{(-1)}}\} - \tan.^{-1}\{\varepsilon^{-\theta\sqrt{(-1)}}\}}{\sqrt{(-1)}} = \log.\tan. \left\{ \left(\frac{\pi}{4}\right) + \left(\frac{\theta}{2}\right) \right\}$$

$$4. -\ \ldots\ldots \tan^{-1}\{x.\varepsilon^{\theta\sqrt{(-1)}}\} + \tan.^{-1}\{x.\varepsilon^{-\theta\sqrt{(-1)}}\} = \tan.^{-1}\{\cos.\theta.\tan.(2\tan^{-1}x)\}$$

$$5. -\ \ldots\ldots \cos.^{-1}\{\varepsilon^{\theta\sqrt{(-1)}}\} + \cos.^{-1}\{\varepsilon^{-\theta\sqrt{(-1)}}\} = \cos.^{-1}\{1 - 2.\sin.\theta\}$$

$$6. -\ \ldots\ldots \cos.^{-1}\{\varepsilon^{\theta\sqrt{(-1)}}\} - \cos.^{-1}\{\varepsilon^{-\theta\sqrt{(-1)}}\} = \cos.^{-1}\{1 + 2.\sin.\theta\}$$

$$= \frac{1}{\sqrt{(-1)}}\log. \left\{ 1 + 2.(2\sin.\theta)^{\frac{1}{2}}.\cos.\left(\frac{\pi}{4} - \frac{\theta}{2}\right) + (2.\sin.\theta) \right\}$$

We will now proceed to the derivation of some theorems, from which many curious results may be deduced.

$$\text{Let } f(x) - A_0 = A_1 x + A_2 x^2 + \&c.$$

For x write successively

$$\left. \begin{matrix} x.\varepsilon^{\theta\sqrt{(-1)}} \\ x.\varepsilon^{-\theta\sqrt{(-1)}} \end{matrix} \right\}, \quad \left. \begin{matrix} x.\varepsilon^{2\theta\sqrt{(-1)}} \\ x.\varepsilon^{-2\theta\sqrt{(-1)}} \end{matrix} \right\}, \ldots \&c.$$

and let the *sums* of the corresponding results be added together, and the whole divided by 2; thus

$$\left\{ \frac{f\{x.\varepsilon^{\theta\sqrt{(-1)}}\} + f\{x.\varepsilon^{-\theta\sqrt{(-1)}}\}}{2} - A_0 \right\} + \&c. =$$

$$= \frac{1}{2} \left\{ A_1.x \left(\frac{\varepsilon^{\theta\sqrt{(-1)}}}{1 - \varepsilon^{\theta\sqrt{(-1)}}} + \frac{\varepsilon^{-\theta\sqrt{(-1)}}}{1 - \varepsilon^{-\theta\sqrt{(-1)}}} \right) + \&c. \right.$$

Suppose u_i any function of i; we shall denote by $S\{u_i\}$ the series $u_1 + u_2 +$ &c. ad inf. and in general by $S\{u_i\},\ \begin{bmatrix} i = m \\ i = n \end{bmatrix}$, the series $u_{m+1} + u_{m+2} + \ldots u_n$ (See Note 2.)

Thus we have

$$S\left\{\frac{f\{x.\varepsilon^{i\theta\sqrt{(-1)}}\} + f\{x.\varepsilon^{-i\theta\sqrt{(-1)}}\}}{2} - A_0\right\} =$$

$$= \frac{1}{2} S\left\{A_i x^i . \left(\frac{\varepsilon^{i\theta\sqrt{(-1)}}}{1 - \varepsilon^{i\theta\sqrt{(-1)}}} + \frac{\varepsilon^{-i\theta\sqrt{(-1)}}}{1 - \varepsilon^{-i\theta\sqrt{(-1)}}}\right)\right\} =$$

$$= \frac{1}{2} S\left\{ - A_i x^i\right\} = \frac{-1}{2} . \left\{f(x) - A_0\right\} \ldots\ldots\ldots\ldots \{1\}$$

Let the nth derived function of both sides be taken, with respect to x, and for $f\{x\}$ substitute $D^{-n}f(x)$; we have

$$S\left\{\varepsilon^{i.n\theta\sqrt{(-1)}} f\{x.\varepsilon^{i\theta\sqrt{(-1)}}\} + \varepsilon^{-i.n\theta\sqrt{(-1)}} f\{x.\varepsilon^{-i\theta\sqrt{(-1)}}\}\right\} = -f(x) .. \{1,1\}$$

(See Note 3.)

Secondly, for x let us write successively

$$\left.\begin{array}{c} x.\varepsilon^{2\theta\sqrt{(-1)}} \\ x.\varepsilon^{-2\theta\sqrt{(-1)}} \end{array}\right\}, \quad \left.\begin{array}{c} x.\varepsilon^{4\theta\sqrt{(-1)}} \\ x.\varepsilon^{-4\theta\sqrt{(-1)}} \end{array}\right\}, \text{ \&c.}$$

and let the results, after being respectively subtracted, be divided by $2\sqrt{(-1)}$; The aggregate of the whole gives

$$S\left\{\frac{f\{x.\varepsilon^{2i\theta\sqrt{(-1)}}\} - f\{x.\varepsilon^{-2i\theta\sqrt{(-1)}}\}}{\sqrt{(-1)}}\right\} = S\left\{A_i x^i . \cot i\theta\right\} \ldots \{2\}$$

In like manner, by following the operations indicated in the first members, we obtain the equations

$$S\left\{\frac{f\{x.\varepsilon^{(2i-1)\theta\sqrt{(-1)}}\} - f\{x.\varepsilon^{-(2i-1)\theta\sqrt{(-1)}}\}}{\sqrt{(-1)}}\right\} = S\left\{A_i x^i . \frac{1}{\sin i\theta}\right\} .. \{3\}$$

$$4 . S\left\{\frac{f\{x.\varepsilon^{(4i-3)\theta\sqrt{(-1)}}\} + f\{x.\varepsilon^{-(4i-3)\theta\sqrt{(-1)}}\}}{2} - A_0\right\} = S\left\{A_i \frac{x^i}{\cos i\theta}\right\} .. \{4\}$$

By introducing $\sin i\theta$ into the second member of this we should obtain an expression of the same kind for $S\{A_i x^i . \tan i\theta\}$, but it may also be done as follows,

$$\frac{1}{\sin. 2i\theta} - \text{cotan}. 2i\theta = \tan. i\theta,$$

In the equations {3} and {2} therefore, for θ write 2θ, and subtracting, we have

$$S\left\{(-1)^{i+1}\frac{f\{x.\varepsilon^{2i\theta\sqrt{(-1)}}\} - f\{x.\varepsilon^{-2i\theta\sqrt{(-1)}}\}}{\sqrt{(-1)}}\right\} = S\{A_i x^i. \tan. i\theta\} \ldots \{5\}$$

Let us now, instead of continuing our geometric series to infinity, take their sums to m terms only, and instead of equation {1}, we have

$$S\left\{\frac{f\{x.\varepsilon^{2i\theta\sqrt{(-1)}}\} + f\{x.\varepsilon^{-2i\theta\sqrt{(-1)}}\}}{2} - A_0\right\} \begin{bmatrix} i = 0 \\ i = m \end{bmatrix} =$$

$$= S\left\{A_i x^i. \frac{\sin. (i.m\theta).\cos. i(m+1)\theta}{\sin. i\theta}\right\} \begin{bmatrix} i = 0 \\ i = \infty \end{bmatrix} \ldots \ldots \ldots \{6\}$$

In like manner,

$$S\left\{\frac{f\{x.\varepsilon^{2i\theta\sqrt{(-1)}}\} - f\{x.\varepsilon^{-2i\theta\sqrt{(-1)}}\}}{2\sqrt{(-1)}}\right\} \begin{bmatrix} i = 0 \\ i = m \end{bmatrix} =$$

$$= S\left\{A_i x^i. \frac{\sin. (i.m\theta).\sin. i(m+1)\theta}{\sin. i\theta}\right\} \begin{bmatrix} i = 0 \\ i = \infty \end{bmatrix} \ldots \ldots \ldots \{7\}$$

$$S\left\{\frac{f\{x.\varepsilon^{(2i-1)\theta\sqrt{(-1)}}\} + f\{x.\varepsilon^{-(2i-1)\theta\sqrt{(-1)}}\}}{2} - A_0\right\} \begin{bmatrix} i = 0 \\ i = m \end{bmatrix} =$$

$$= S\left\{A_i x^i. \frac{\sin. (i.m\theta).\cos. i(m\theta)}{\sin. i\theta}\right\} \begin{bmatrix} i = 0 \\ i = \infty \end{bmatrix} \ldots \ldots \ldots \{8\}$$

$$S\left\{\frac{f\{x.\varepsilon^{(2i-1)\theta\sqrt{(-1)}}\} - f\{x.\varepsilon^{-(2i-1)\theta\sqrt{(-1)}}\}}{2\sqrt{(-1)}}\right\} \begin{bmatrix} i = 0 \\ i = m \end{bmatrix} =$$

$$= S\left\{A_i x^i. \frac{\sin. (i.m\theta).\sin. (i.m\theta)}{\sin. i\theta}\right\} \begin{bmatrix} i = 0 \\ i = \infty \end{bmatrix} \ldots \ldots \ldots \{9\}$$

$$S\left\{\frac{f\{x.\varepsilon^{(4i-3)\theta\sqrt{(-1)}}\} + f\{x.\varepsilon^{-(4i-3)\theta\sqrt{(-1)}}\}}{2} - A_0\right\} \begin{bmatrix} i = 0 \\ i = m \end{bmatrix} =$$

$$= S\left\{A_i x^i. \frac{\sin. (i.2m\theta).\cos. i(2m-1)\theta}{\sin. (i.2\theta)}\right\} \begin{bmatrix} i = 0 \\ i = \infty \end{bmatrix} \ldots \ldots \ldots \{10\}$$

$$S\left\{\frac{f\{x.\varepsilon^{(4i-3)\theta\sqrt{(-1)}}\} - f\{x.\varepsilon^{-(4i-3)\theta\sqrt{(-1)}}\}}{2\sqrt{(-1)}}\right\} \begin{bmatrix} i = 0 \\ i = m \end{bmatrix} =$$

K

$$= S\left\{A_i x^i . \frac{\sin.(i.2m\theta).\sin. i(2m-1)\theta}{\sin.(i.2\theta)}\right\}\quad\begin{bmatrix}i=0\\i=\infty\end{bmatrix}\ldots\ldots\ldots\{11\}$$

By assigning particular values to θ, we shall arrive at the actual sums of several series. Thus the second member of Equation $\{6\}$ becomes

$$\frac{1}{2}S\left\{A_i x^i.(\sin. i.2m\theta.\cotan. i\theta - 1 + \cos. i.2m\theta)\right\}$$

But
$$S\left\{A_i x^i.(1-\cos. i.2m\theta)\right\} =$$

$$= f(x) - A_0 - \left\{\frac{f\{x.\varepsilon^{2m\theta\sqrt{(-1)}}\} + f\{x.\varepsilon^{-2m\theta\sqrt{(-1)}}\}}{2} - A_0\right\}$$

and consequently,

$$S\left\{A_i x^i.\sin. i.2m\theta.\cotan. i\theta\right\}\quad\begin{bmatrix}i=0\\i=\infty\end{bmatrix} =$$

$$= 2.S\left\{\frac{f\{x.\varepsilon^{2i\theta\sqrt{(-1)}}\} + f\{x.\varepsilon^{-2i\theta\sqrt{(-1)}}\}}{2} - A_0\right\}\quad\begin{bmatrix}i=0\\i=m\end{bmatrix}$$

$$+ f\{x\} - \left\{\frac{f\{x.\varepsilon^{2m\theta\sqrt{(-1)}}\} + f\{x.\varepsilon^{-2m\theta\sqrt{(-1)}}\}}{2}\right\}\ldots\ldots\ldots\{6,1\}$$

Suppose now $\theta = \left(\frac{2n+1}{2m}\right).\left(\frac{\pi}{2}\right)$, n being any integer, and for every even value of i, $\sin. i.2m\theta$ vanishes, but for every odd one it becomes

$$(-1)^{\frac{(2n+1).i+3}{2}}$$

for i therefore write $2i-1$, and we find, for these values of θ,

$$S\left\{(-1)^{i+1}.A_{2i-1}.x^{2i-1}\cotan.(2i-1)\theta\right\}\quad\begin{bmatrix}i=0\\i=\infty\end{bmatrix} =$$

$$=(-1)^n.\left\{\begin{array}{l}2.S\left\{\dfrac{f\{x.\varepsilon^{2i\theta\sqrt{(-1)}}\} + f\{x.\varepsilon^{-2i\theta\sqrt{(-1)}}\}}{2} - A_0\right\}\quad\begin{bmatrix}i=0\\i=m\end{bmatrix}\\[6pt]+ f(x)\\[6pt]-\left\{\dfrac{f\{x.\varepsilon^{2m\theta\sqrt{(-1)}}\} + f\{x.\varepsilon^{-2m\theta\sqrt{(-1)}}\}}{2}\right\}\end{array}\right\}..\{6,2\}$$

Let us denote the right hand member of this by Q_x.

Again, from Equation 7; for the same values of θ,

$$S\left\{A_{2i-1}.x^{2i-1}.\cotan.(2i-1)\theta\right\} + 2.S\left\{A_{4i-2} x^{4i-2}.\cotan.(4i-2)\theta\right\} =$$

$$= 2 . S \left\{ \frac{f\{x . \varepsilon^{2i\theta\sqrt{(-1)}}\} - f\{x . \varepsilon^{-2i\theta\sqrt{(-1)}}\}}{2\sqrt{(-1)}} \right\} \begin{bmatrix} i = 0 \\ i = m \end{bmatrix} - \left\{ \frac{f\{x . \varepsilon^{2m\theta\sqrt{(-1)}}\} - f\{x . \varepsilon^{-2m\theta\sqrt{(-1)}}\}}{2\sqrt{(-1)}} \right\}$$

Let this be called R_x, We have then

$$Q_x + R_x \text{ or } (Q + R)_x = \qquad \text{(See Note 4.)}$$

$$2 . S \left\{ A_{4i-2} \, x^{4i-2} . \cot. (4i - 2)\theta \right\} + S \left\{ (1 + (-1)^{i+1}) A_{2i-1} \, x^{2i-1} . \cot. (2i - 1)\theta \right\}$$

or,

$$S \left\{ A_{4i-3} \, x^{4i-3} . \cot. (4i - 3)\theta + A_{4i-2} \, x^{4i-2} . \cot. (4i - 2)\theta \right\} = \frac{1}{2} (Q + R)_x . . \{6, 3\}$$

In like manner,

$$S \left\{ A_{4i-2} \, x^{4i-2} . \cot. (4i - 2)\theta + A_{4i-1} \, x^{4i-1} . \cot. (4i - 1)\theta \right\} = \frac{1}{2} (R - Q)_x . . \{6, 4\}$$

From these, by performing the operations indicated as follows, we find,

$$\frac{1}{4} \left\{ (Q + R)_x + (Q + R)_{-x} \right\} = S \left\{ A_{4i-2} \, x^{4i-2} . \cotan. (4i - 2)\theta \right\} \ldots \ldots \{6, 5\}$$

$$\frac{1}{4} \left\{ (Q + R)_x - (Q + R)_{-x} \right\} = S \left\{ A_{4i-3} \, x^{4i-3} . \cotan. (4i - 3)\theta \right\} \ldots \ldots \{6, 6\}$$

$$\frac{1}{4} \left\{ (R - Q)_x - (R - Q)_{-x} \right\} = S \left\{ A_{4i-1} \, x^{4i-1} . \cotan. (4i - 1)\theta \right\} \ldots \ldots \{6, 7\}$$

$$\frac{1}{2} \left\{ R_x - R_{-x} \right\} \ldots \ldots \ldots \ldots = S \left\{ A_{2i-1} \, x^{2i-1} . \cotan. (2i - 1)\theta \right\} \ldots . . \{6, 8\}$$

$$Q_x \ldots \ldots \ldots \ldots \ldots \ldots = S \left\{ (-1)^{i+1} . A_{2i-1} \, x^{2i-1} . \cotan. (2i - 1)\theta \right\}$$

From the equation $\{9\}$, whose second member $= S \left\{ A_i x^i . \frac{1 - \cos. 2 i m\theta}{2 . \sin. i\theta} \right\}$ we find,

$$S \left\{ A_{2i-1} \frac{x^{2i-1}}{\sin. (2i - 1)\theta} \right\} + 2 . S \left\{ A_{4i-2} \frac{x^{4i-2}}{\sin. (4i - 2)\theta} \right\} =$$

$$= 2 S \left\{ \frac{f\{x . \varepsilon^{(2i-1)\theta\sqrt{(-1)}}\} - f\{x . \varepsilon^{-(2i-1)\theta\sqrt{(-1)}}\}}{2\sqrt{(-1)}} \right\} \begin{bmatrix} i = 0 \\ i = m \end{bmatrix}$$

Let us for a moment call this K_x; and we have

$$\frac{1}{4} (K_x + K_{-x}) = S \left\{ A_{4i-2} \frac{x^{4i-2}}{\sin. (4i - 2)\theta} \right\} \ldots \ldots \ldots \ldots \{9, 1\}$$

$$\frac{1}{2} (K_x - K_{-x}) = S \left\{ A_{2i-1} \frac{x^{2i-1}}{\sin. (2i - 1)\theta} \right\} \ldots \ldots \ldots \ldots \{9, 2\}$$

In equation 10, for θ, writing $\left(\dfrac{2n+1}{2m}\right) \cdot \left(\dfrac{\pi}{2}\right)$ we find at once,

$$S\left\{A_{2i-1}\,\frac{x^{2i-1}}{\cos.\,(2i-1)\,\theta}\right\} =$$

$$= 2\,.\,S\left\{\frac{f\{x\,.\,\varepsilon^{(4i-3)\theta\sqrt{(-1)}}\} + f\{x\,.\,\varepsilon^{-(4i-3)\theta\sqrt{(-1)}}\}}{2} - A_0\right\} \begin{bmatrix} i = 0 \\ i = m \end{bmatrix} \cdot\cdot \{10,1\}$$

Let the equation $\{6,8\}$, be subtracted from $\{9,2\}$, and we get

$$\frac{1}{2}\,((K-R)_x - (K-R)_{-x}) = S\left\{A_{2i-1}\,x^{2i-1}.\,\tan.\,(2i-1)\cdot\frac{\theta}{2}\right\} \ldots\ldots \{6,9;1\}$$

In like manner, by subtracting $\{6,5\}$ from $\{9,1\}$, we get

$$\frac{1}{4}\left\{(K-Q-R)_x + (K-Q-R)_{-x}\right\} = S\left\{A_{4i-2}\,x^{4i-2}.\,\tan.\,(4i-2)\frac{\theta}{2}\right\} \cdot\cdot \{6,9;2\}$$

Next, let $\theta = \left(\dfrac{2n+1}{m}\right) \cdot \left(\dfrac{\pi}{2}\right)$. And we find, in the same way,

$$S\left\{A_{2i-1}\,x^{2i-1}.\,\cot.\,(2i-1)\theta\right\} =$$

$$= \left\{ \begin{aligned} &S\left\{\frac{f\{x\,.\,\varepsilon^{2i\theta\sqrt{(-1)}}\} - f\{x\,.\,\varepsilon^{-2i\theta\sqrt{(-1)}}\}}{2\sqrt{(-1)}}\right\} \begin{bmatrix} i = 0 \\ i = m \end{bmatrix} \\ &-\frac{1}{2}\cdot\left\{\frac{f\{x\,.\,\varepsilon^{2m\theta\sqrt{(-1)}}\} - f\{x\,.\,\varepsilon^{-2m\theta\sqrt{(-1)}}\}}{2\sqrt{(-1)}}\right\} \end{aligned} \right\} \ldots\ldots \{7,1\}$$

and,
$$S\left\{A_{2i-1}\,\frac{x^{2i-1}}{\sin.\,(2i-1)\,\theta}\right\} =$$

$$= S\left\{\frac{f\{x\,.\,\varepsilon^{(2i-1)\theta\sqrt{(-1)}}\} - f\{x\,.\,\varepsilon^{-(2i-1)\theta\sqrt{(-1)}}\}}{2\sqrt{(-1)}}\right\} \begin{bmatrix} i = 0 \\ i = m \end{bmatrix} \ldots \{9,3\}$$

Lastly, let $\theta = \dfrac{2n+1}{4m}\cdot\left(\dfrac{\pi}{2}\right)$. And equation $\{10\}$ gives, if we put

$$T_x \text{ for } 4\,.\,S\left\{\frac{f\{x\,.\,\varepsilon^{(4i-3)\sqrt{(-1)}\,.\,\theta}\} + f\{x\,.\,\varepsilon^{-(4i-3)\theta\sqrt{(-1)}}\}}{2} - A_0\right\} \begin{bmatrix} i - 0 \\ i = m \end{bmatrix}$$

$$\frac{(-1)^n}{2}\cdot(T_x - T_{-x}) = S\left\{A_{2i-1}\,x^{2i-1}\left(\frac{(-1)^{i+1}}{\sin.\,(2i-1)\,\theta} + \frac{(-1)^n}{\cos.\,(2i-1)\,\theta}\right)\right\} \ldots \{10,2\}$$

and,
$$\frac{1}{4}\cdot(T_x + T_{-x}) = S\left\{A_{4i-2}\,\frac{x^{4i-2}}{\cos.\,(4i-2)\,\theta}\right\} \ldots\ldots \{10,3\}$$

In like manner, if we call the function

$$4. \quad S \left\{ \frac{f\left\{x.\varepsilon^{(4i-3)\theta\sqrt{(-1)}}\right\} - f\left\{x.\varepsilon^{-(4i-3)\theta\sqrt{(-1)}}\right\}}{2.\sqrt{(-1)}} \right\} \quad \begin{bmatrix} i=0 \\ i=m \end{bmatrix}$$

U_x, we shall find,

$$\frac{(-1)^{n+1}}{2}.(U_x - U_{-x}) = S \left\{ A_{2i-1} x^{2i-1} \left(\frac{(-1)^{i+1}}{\cos.(2i-1)\theta} - \frac{(-1)^n}{\sin.(2i-1)\theta} \right) \right\} .. \{11,1\}$$

$$\frac{1}{4} \left\{ U_x + U_{-x} \right\} = S \left\{ A_{4i-2} . \frac{x^{4i-2}}{\sin.(4i-2)\theta} \right\} \dots\dots \{11,2\}$$

We have not as yet assigned particular values to $f(x)$; We will now therefore select some of the most remarkable forms, principally to illustrate the application of Equation $\{1\}$, which indeed affords an inexhaustible store of curious results. The other Equations will be useful in transforming certain series into infinite products.

1.   Let $f(x) = \tan.^{-1}x = \dfrac{x}{1} - \dfrac{x^3}{3} + \&c.$

The Equation $\{1\}$ by the help of the 4th of the formulæ in page (35), gives

$$S \left\{ \tan.^{-1} \left(\frac{2x}{1-x^2} . \cos. i\theta \right) \right\} = - \tan.^{-1} x$$

or, if $x = \tan. \dfrac{1}{2} A$.

$$\tan.^{-1} (\tan. A \cos.\theta) + \tan.^{-1} (\tan. A \cos. 2\theta) + \tan.^{-1} (\tan. A \cos. 3\theta) + \&c. = - \frac{A}{2}$$

or, which is the same thing,

$$P \left\{ \frac{1+\sqrt{(-1)}.\tan. A . \cos. i\theta}{1-\sqrt{(-1)}.\tan. A . \cos. i\theta} \right\} = \varepsilon^{\frac{A}{\sqrt{(-1)}}}$$

where $P \{u_i\} = u_1.u_2 \dots u_\infty$,   and $P \{u_i\} \begin{bmatrix} i=m \\ i=n \end{bmatrix} = u_{m+1} \dots u_n$

Let now $\sqrt{(-1)}.\tan. A = a$, and we find

$$P \left\{ \frac{1+a . \cos. i\theta}{1-a . \cos. i\theta} \right\} = \left(\frac{1-a}{1+a} \right)^{\frac{1}{2}} \dots\dots \{1,3\}$$

If $A = \dfrac{\pi}{4}$, we have

$$\tan.^{-1} \cos. \theta + \tan.^{-1} \cos. 2\theta + \tan.^{-1} \cos. 3\theta + \&c. = - \frac{\pi}{8} \dots \{1,4\}$$

The equation $\{1,1\}$ gives us

$$-\tan.^{-1}x = S\left\{\varepsilon^{i.n\theta\sqrt{(-1)}}.\tan.^{-1}(x.\varepsilon^{i\theta\sqrt{(-1)}}) + \varepsilon^{-in\theta\sqrt{(-1)}}.\tan.^{-1}(x.\varepsilon^{-i\theta\sqrt{(-1)}})\right\}$$

$$= S\left\{\cos.in\theta\{\tan.^{-1}(x.\varepsilon^{i\theta\sqrt{(-1)}}) + \tan.^{-1}(x.\varepsilon^{-i\theta\sqrt{(-1)}})\} - \sin.in\theta.\frac{\tan.^{-1}(x.\varepsilon^{i\theta\sqrt{(-1)}}) - \tan.^{-1}(x.\varepsilon^{-i\theta\sqrt{(-1)}})}{\sqrt{(-1)}}\right\}$$

$$= S\left\{\cos.in\theta.\tan.^{-1}\left(\frac{2x\cos.i\theta}{1-x^2}\right) - \sin.in\theta.\frac{1}{2}\log.\left(\frac{1+2x.\sin.i\theta+x^2}{1-2x.\sin.i\theta+x^2}\right)\right\}$$

Let $x=1$, and we have

$$-\frac{\pi}{4} = S\left\{\frac{\pi}{2}.\cos.in\theta - \sin.in\theta.\log.\left\{\tan.\left(\frac{\pi}{4}+\frac{i\theta}{2}\right)\right\}\right\}$$

Now, $S\left\{\frac{\pi}{2}.\cos.in\theta\right\} = \frac{\pi}{2}.S(\cos.in\theta) = -\frac{\pi}{4}$; thus,

$$0 = S\left\{\log.\left(\tan.\left(\frac{\pi}{4}+\frac{i\theta}{2}\right)\right)^{\sin.in\theta}\right\}$$

Whence, writing 2θ for θ, we get

$$1 = P\left\{\left(\tan.\left(\frac{\pi}{4}+i\theta\right)\right)^{\sin.2in\theta}\right\}\ldots.\{1,5\}$$

if $n=\frac{1}{2}$, (See Note 3)

$$1 = P\left\{\left(\tan.\left\{\frac{\pi}{4}+i\theta\right\}\right)^{\sin.i\theta}\right\} =$$

$$= \left\{\tan.\left(\frac{\pi}{4}+\theta\right)\right\}^{\sin.\theta}.\left\{\tan.\left(\frac{\pi}{4}+2\theta\right)\right\}^{\sin.2\theta}\left\{\tan.\left(\frac{\pi}{4}+3\theta\right)\right\}^{\sin.3\theta}.\&c\ldots\{1,6\}$$

Analogous to this is a theorem of so singular a form, that we shall retard our immediate progress to give it insertion, though derived from principles somewhat more extended;

$$\frac{1}{1}\tan.^{-1}x = \frac{x}{1.1} - \frac{x^3}{3.1} + \frac{x^5}{5.1} - \&c. \left.\begin{array}{c}\\\\\\\end{array}\right\}$$

$$\frac{1}{3}\tan.^{-1}(x^3) = \frac{x^3}{1.3} - \frac{x^9}{3.3} + \frac{x^{15}}{5.3} - \&c. \left.\begin{array}{c}\\\\\\\end{array}\right\} \text{Adding vertically}$$

$$\frac{1}{5}\tan.^{-1}(x^5) = \frac{x^5}{1.5} - \frac{x^{15}}{3.5} + \frac{x^{2b}}{5.5} - \&c. \left.\begin{array}{c}\\\\\\\end{array}\right\}$$

$$S\left\{\frac{\tan.^{-1}(x^{2i-1})}{2i-1}\right\} = \frac{1}{2}\left\{\log.\left(\frac{1+x}{1-x}\right)^{\frac{1}{1}} - \log.\left(\frac{1+x^3}{1-x^3}\right)^{\frac{1}{3}} + \&c.\right\} =$$

$$= \frac{1}{2}\ S\left\{\log.\left(\frac{1+x^{2i-1}}{1-x^{2i-1}}\right)^{\frac{(-1)^{i+1}}{2i-1}}\right\}$$

In like manner, writing x^{-1} for x,

$$S\left\{\frac{\tan.^{-1}(x.^{-(2i-1)})}{2i-1}\right\} = \frac{1}{2}\ S\left\{\log.\left(\frac{1+x^{-(2i-1)}}{1-x^{-2i-1}}\right)^{\frac{(-1)^{i+1}}{2i-1}}\right\} \dots\dots \{12\}$$

and subtracting

$$S\left\{\frac{\tan.^{-1}(x^{2i-1}) - \tan.^{-1}(x^{-(2i-1)})}{2i-1}\right\} = \frac{1}{2}\ S\left\{\log.(-1)^{\frac{(-1)^{i+1}}{2i-1}}\right\}$$

$$= \frac{1}{2}\ S\left\{\frac{(-1)^{i+1}}{2i-1}.\log.(-1)\right\} = \frac{1}{2}\log.(-1).S\left\{\frac{(-1)^{i+1}}{2i-1}\right\} =$$

$$= \frac{\sqrt{(-1)}}{2}.\pi.\frac{\pi}{4} = \frac{\pi^2}{8}.\sqrt{(-1)}.$$

Now, let $x = \varepsilon^{2\theta\sqrt{(-1)}}$, and we get

$$S\left\{\frac{1}{2i-1}.\frac{\tan.^{-1}(\varepsilon^{(4i-2)\theta\sqrt{(-1)}}) - \tan.^{-1}(\varepsilon^{-(4i-2)\theta\sqrt{(-1)}})}{\sqrt{(-1)}}\right\} = \frac{\pi^2}{8}, \quad \text{or,}$$

$$\varepsilon^{\frac{\pi^2}{8}} = P\left\{\left(\tan.\left\{\frac{\pi}{4}+(2i-1)\theta\right\}\right)^{\frac{1}{2i-1}}\right\}. - \text{and for } \theta \text{ writing } \left(\frac{\pi}{4}+\theta\right)$$

$$\varepsilon^{-\frac{\pi^2}{8}} = P\left\{\left(\tan.\left\{\frac{i\pi}{2}+(2i-1)\theta\right\}\right)^{\frac{-1}{2i-1}}\right\} =$$

$$= (\tan.\theta)^{\frac{1}{1}}.(\cot.3\theta)^{\frac{1}{3}}.(\tan.5\theta)^{\frac{1}{5}}.\&c.\dots\dots \{12,1\}$$

To return,

By differentiating $\{1,3\}$, with respect to a, we find

$$S\left\{\frac{\cos.i\theta}{1-(a.\cos.i\theta)^2}\right\} = -\frac{1}{2(1-a^2)}\dots\dots\{1,7\}$$

The application of equation $\{2\}$, and the third of the formulæ in page (35), gives

$$P\left\{\left(\frac{\frac{x^{4i-3}}{4i-3}\cdot\text{cotan.}\,(4i-3)\,\theta}{\frac{x^{4i-1}}{\varepsilon^{4i-1}}\cdot\text{cotan.}\,(4i-1)\,\theta}\right)^2\right\} = P\left\{\frac{1+2\,x\cdot\sin.\,(2\,i\theta)+x^2}{1-2\,x\cdot\sin.\,(2\,i\theta)+x^2}\right\}\ldots\{2,1\}$$

If $x = 1$, we obtain, writing $\frac{\pi}{2} - \theta$ for θ,

$$P\left\{\frac{\varepsilon^{\frac{\tan.\,(4i-3)\,\theta}{4i-3}}}{\varepsilon^{\frac{\tan.\,(4i-1)\,\theta}{4i-1}}}\right\} = P\left\{\frac{\tan.\left(2\,i\,\theta-\frac{\pi}{4}\right)}{\tan.\left((2\,i-1)\,\theta-\frac{\pi}{4}\right)}\right\} =$$

$$= P\left\{\frac{\cos.\,(4\,i-1)\,\theta-\sin.\,\theta}{\cos.\,(4\,i-1)\,\theta+\sin.\,\theta}\right\}\ldots\ldots\{2,2\}$$

in virtue of the formula,

$$\frac{\tan.\,A}{\tan.\,B} = \frac{\sin.\,(A+B)+\sin.\,(A-B)}{\sin.\,(A+B)-\sin.\,(A-B)}$$

Let us next suppose $f(x) = \varepsilon^x$, and we have

$$S\left\{\varepsilon^{x\,.\,\cos.\,i\,\theta}\ldots\ldots\cos.\,(x\sin.\,i\,\theta)-1\right\} = -\frac{1}{2}\{\varepsilon^x-1\}\ldots\ldots\{1,8\}$$

and, by the equation $\{1,1\}$, we find, making $n\theta = \phi$

$$S\left\{\varepsilon^{x\,.\,\cos.\,i\,\theta}\ldots\ldots\cos.\,\{i\phi+x\,.\,\sin.\,i\,\theta\}\right\} = -\frac{\varepsilon^x}{2}\ldots\ldots\ldots\{1,9\}$$

and if $x = 1$,

$$S\left\{\varepsilon^{\cos.\,i\,\theta}\,.\,\cos.\,(i\phi+\sin.\,i\,\theta)\right\} = -\frac{\varepsilon}{2}\ldots\ldots\ldots\ldots\{1,10\}$$

It were easy to multiply examples, but as most of the equations derived from $\{1\}$ carry somewhat of indistinctness with them to the imagination, we shall confine ourselves to one more, viz.

$$P\left\{1-2\,x\,.\,\cos.\,i\,\theta+x^2\right\} = \frac{1}{1-x}\ldots\ldots\{1,11\}$$

which is easily obtained, by making $f(x) = -\log.\,(1-x)$.

Examples of the actual summation of series of tangents will be given hereafter (See Note 5). Meanwhile, we have seen that the foregoing summations depend entirely on the commensurability of θ with π; when this condition holds, the coefficient tan. θ, tan. 2θ, &c. recur after m terms in the same order; and thus it happens, that the sums of these series consist of as many terms as there are units in m, the denominator of $\dfrac{\theta}{\pi}$ or $\dfrac{\theta}{\frac{1}{2}\pi}$. In other cases we may without hesitation affirm, that the summation of such series surpasses the powers of analysis. In fact, the same formulæ which represent their sums, would enable us to assign the sums of such series, as

$$f(x) + f(x^2) + \ldots f(x^n)$$

$$\left\{ f\{xv\} - A_0 \right\} + \left\{ f\{xv^2\} - A_0 \right\} + \text{&c. ad inf.}$$

To shew this connexion, we have

$$\left. \begin{aligned} f(xv) - A_0 &= A_1 \, xv + A_2 \, x^2 v^2 + \text{&c.} \\ f(xv^2) - A_0 &= A_1 \, xv^2 + A_2 \, x^2 v^4 + \text{&c.} \\ \text{&c.} &= \quad \text{&c.} \end{aligned} \right\} \quad \text{and adding}$$

$$S\left\{ f(xv^i) - A_0 \right\} = S\left\{ A_i \, x^i . \frac{v^i}{1-v^i} \right\} = S\left\{ A_i \, x^i . \frac{-v^{\frac{i}{2}}}{v^{\frac{i}{2}} - v^{-\frac{i}{2}}} \right\}$$

$$= - \, S\left\{ A_i \, x^i . \frac{\varepsilon^{\frac{i.\log. v}{2\sqrt{(-1)}} \sqrt{(-1)}}}{\varepsilon^{\frac{i.\log. v}{2\sqrt{(-1)}} \sqrt{(-1)}} - \varepsilon^{\frac{-i.\log. v}{2\sqrt{(-1)}} \sqrt{(-1)}}} \right\} =$$

$$= - \, S\left\{ \frac{A_i \, x^i}{2\sqrt{(-1)}} . \left(\frac{\cos.\dfrac{i.\log. v}{2\sqrt{(-1)}} + \sqrt{(-1)} . \sin.\dfrac{i.\log. v}{2\sqrt{(-1)}}}{\sin.\dfrac{i.\log. v}{2\sqrt{(-1)}}} \right) \right\} =$$

$$= - \, S\left\{ \frac{A_i \, x^i}{2} + \frac{A_i \, x^i}{2\sqrt{(-1)}} . \cot. \, i \left(\frac{\log. v}{2\sqrt{(-1)}} \right) \right\} =$$

$$= - \, \frac{1}{2} \left\{ f(x) - A_0 \right\} - \frac{1}{2\sqrt{(-1)}} . S\left\{ A_i \, x^i . \cot. \, i \left(\frac{\log. v}{2\sqrt{(-1)}} \right) \right\}$$

For x, write xv^n, and we have

$$S\left\{ f(xv^{n+i}) - A_0 \right\} =$$

$$= - \, \frac{1}{2} \left(f(xv^n) - A_0 \right) - \frac{1}{2\sqrt{(-1)}} . S\left\{ A_i \, (xv^n)^i . \cot. \, i \left(\frac{\log. v}{2\sqrt{(-1)}} \right) \right\}$$

M

and
$$S\left\{f(xv^i) - A_0\right\} =$$

$$= -\frac{1}{2}\left\{f(x) - A_0\right\} - \frac{1}{2\sqrt{(-1)}} S\left\{A_i x^i . \cot. i \left(\frac{\log.. v}{2\sqrt{(-1)}}\right)\right\}$$

and subtracting,

$$S\left\{f(xv^i) - A_0\right\} \begin{bmatrix} i=0 \\ i=n \end{bmatrix} =$$

$$= -\frac{1}{2}\left\{f(xv^n) - f(x)\right\} - \frac{1}{2\sqrt{(-1)}} . S\left\{A_i x^i (v^{ni} - 1) . \cot. i . \left(\frac{\log. v}{2\sqrt{(-1)}}\right)\right\}$$

Here then we shall conclude the more connected part of our remarks; contented with having exhibited in a tangible, and not inelegant shape, some fragments of a theory as yet waste and barren, without laying claim to the praise of very profound investigation, or even of absolute originality. What farther we have to offer on the subject, being of a miscellaneous complexion, we shall prefer giving in the form of a note.   See Note 5.

NOTES.

NOTE I.

$f^0 x = x.$ This equation may be considered as the characteristic mark of the correctness of the notation we have adopted for the successive functions of a symbol x. Unless this condition hold, any equation, such as

$$f^n(x) = \phi\{n, x\}$$

will not continue true for the negative values of n. We may here remark, that the notation adopted by the Author of the Logarithmic Transcendants, is calculated in some degree to mislead the mind into a supposition that the successive logarithms of x are represented by the symbols $\mathbf{L}^1(x)$, $\mathbf{L}^2(x)$, &c. Now if we consider the value of $\mathbf{L}^0(x)$, we shall find it equal to $\dfrac{-1+x}{x}$, instead of x. The criterion here mentioned does not therefore hold in this instance.

When we consider the nth function (f^n) of a symbol x as a function of two symbols, n and x, it becomes interesting to obtain an expression for it in the form above mentioned.

$$f^n(x) = \phi\{n, x\},$$

Now in some simple instances this may be accomplished by mere substitution; for example,

$$f(x) = \frac{a\,x}{b + c\,x}$$

$$f^2(x) = ff(x) = \frac{a \cdot \dfrac{ax}{b+cx}}{b + c \cdot \dfrac{ax}{b+cx}} = \frac{a^2 x}{b^2 + cx\,(a+b)}$$

$$f^3 x = ff^2 x = \frac{a \cdot \dfrac{a^2 x}{b^2 + cx\,(a+b)}}{b + c \cdot \dfrac{a^2 x}{b^2 + cx\,(a+b)}} = \frac{a^3 x}{b^3 + cx\,(a^2 + ab + b^2)}$$

and so on, till we arrive at the equation

$$f^n(x) = \frac{a^n \cdot x}{b^n + c\,x \left\{\dfrac{a^n - b^n}{a - b}\right\}};$$

but, in general, such an operation would run out into excessive complexity, and that, without bringing us at last to the desired end; we will therefore consider the subject in another light.

The nature of the operation f being given, and as we know that $f^{z+1}(x) = ff^z(x)$, if we suppose $f^z(x) = u_z$, we have an equation of finite differences,

$$0 = u_{z+1} - f(u_z), \ldots \ldots \ldots (a)$$

of the first order, for determining the nature of the function of z which u_z represents. Now in this expression the symbol x will not be *explicitly* found, but $u_z = f^z(x)$, is a function of x. Moreover in the substitution of $f^z(x)$ for x in the expression of $f(x)$, the symbol x remains unaltered. The only way in which x can enter into the constitution of u_z, so as to fulfil this condition, and remain at the same time arbitrary, is by means of the indeterminate constant introduced by the integration of (a). This will be evident also from the following consideration;

Since $u_z = f(u_{z-1})$,   we have, $u_{z-1} = f^{-1}u_z$.

thus, $u_0 = f^{-1}u_1 = f^{-1}\{f(x)\}$ = that symbol whose function (f)

is $f(x)$ or, $u_0 = x$.

Now let C be the arbitrary constant in the integral of (a).

we have, $u_z = F\{z, C\}$,   consequently $u_0 = F(0, C)$

so that C is a function of u_0 or x. We may here remark, that if we can procure only a particular integral, or solution of (a), we shall be enabled to assign the form of $f^z(x)$ only for a particular value of x, viz. the value which u_0 takes in that case. To illustrate this theory by an instance, let

$$f(x) = \frac{\alpha + \beta x}{\gamma + \delta x},$$

the general rational form of the first degree in x.

We have then,

$$u_{z+1} = f(u_z) = \frac{\alpha + \beta \cdot u_z}{\gamma + \delta \cdot u_z},$$

or, $$0 = u_{z+1} \cdot u_z + \left(\frac{\gamma}{\delta}\right) \cdot u_{z+1} - \left(\frac{\beta}{\delta}\right) \cdot u_z - \left(\frac{\alpha}{\delta}\right) \ldots \ldots (b)$$

The complete integral of this equation may be obtained as follows,

Assume $$u_z = \frac{A_z + B_z \cdot H}{C_z + D_z \cdot H} \cdots \cdots \cdots (c)$$

H being an arbitrary constant, and we find

$$\frac{A_{z+1}+B_{z+1}.H}{C_{z+1}+D_{z+1}.H} = u_{z+1} = \frac{a+\beta \cdot \dfrac{A_z+B_z.H}{C_z+D_z.H}}{\gamma+\delta \cdot \dfrac{A_z+B_z.H}{C_z+D_z.H}} = \frac{\{aC_z+\beta A_z\}+H.\{aD_z+\beta B_z\}}{\{\gamma C_z+\delta A_z\}+H.\{\gamma D_z+\delta B_z\}}$$

And equating similar terms, in order that H may remain arbitrary (so as to obtain a complete solution) we get four equations,

$$\left.\begin{array}{l} 0 = A_{z+1} - aC_z - \beta A_z \\ 0 = C_{z+1} - \gamma C_z - \delta A_z \end{array}\right\} \quad \text{and} \quad \left\{\begin{array}{l} 0 = B_{z+1} - aD_z - \beta B_z \\ 0 = D_{z+1} - \gamma D_z - \delta B_z \end{array}\right.$$

whence by elimination,

$$\left.\begin{array}{l} 0 = C_{z+2} - (\beta+\gamma).C_{z+1}+(\beta\gamma-a\delta).C_z \\ 0 = D_{z+2} - (\beta+\gamma).D_{z+1}+(\beta\gamma-a\delta).D_z \end{array}\right\} \quad \text{and} \quad \left\{\begin{array}{l} A_z = \frac{1}{\delta}(C_{z+1}-\gamma.C_z) \\ B_z = \frac{1}{\delta}(D_{z+1}-\gamma.D_z) \end{array}\right.$$

From these, integrating, and writing

$$\varpi = \frac{\beta-\gamma}{2};\dots\lambda = \frac{\beta+\gamma}{2};\dots\mu = \left\{\left(\frac{\beta-\gamma}{2}\right)^2+a\delta\right\}^{\frac{1}{2}};\dots\nu = \left(\frac{\lambda-\mu}{\lambda+\mu}\right);$$

we obtain

$$C_z = h.(\lambda+\mu)^z + k.(\lambda-\mu)^z$$

$$D_z = f.(\lambda+\mu)^z + g.(\lambda-\mu)^z$$

$$A_z = \frac{1}{\delta}\left\{h.(\lambda+\mu)^z.(\lambda+\mu-\gamma) + k.(\lambda-\mu)^z.(\lambda-\mu-\gamma)\right\}$$

$$B_z = \frac{1}{\delta}\left\{f.(\lambda+\mu)^z.(\lambda+\mu-\gamma) + g.(\lambda-\mu)^z.(\lambda-\mu-\gamma)\right\}$$

h, k, f, and g being four arbitrary constants. These values being substituted in the expression (c), it becomes

$$u_z = \frac{1}{\delta} \cdot \frac{(\lambda+\mu-\gamma)+(\lambda-\mu-\gamma).\nu^z.\left\{\dfrac{k+g.H}{h+f.H}\right\}}{1+\nu^z.\left\{\dfrac{k+g.H}{h+f.H}\right\}} =$$

$$= \frac{1}{\delta} \cdot \frac{(\lambda+\mu-\gamma)+(\lambda-\mu-\gamma).\nu^z.K}{1+\nu^z.K} \dots\dots\dots (d)$$

N

Thus we see that the five arbitrary constants H, f, g, h, k, reduce themselves to one, $\dfrac{k+g \cdot H}{h+f \cdot H} = K$, in the expression of u_z.

Let us now see what function K is of x. For this purpose, let $z = 0$, and

$$u_0 = x = \frac{1}{\delta} \cdot \frac{(\lambda + \mu - \gamma) + (\lambda - \mu - \gamma) \cdot K}{1 + K}$$

whence we conclude,

$$K = \frac{\delta \cdot x - (\lambda + \mu - \gamma)}{(\lambda - \mu - \gamma) - \delta \cdot x}$$

which substituted in (d) gives, after all reductions,

$$u_z = f^z(x) = \frac{\{a + (\varpi - \mu) \cdot x\} \cdot \nu^z - \{a + (\varpi + \mu) \cdot x\}}{\{\delta \cdot x - (\varpi + \mu)\} \cdot \nu^z - \{\delta \cdot x - (\varpi - \mu)\}}$$

which, it must be observed, answers equally for negative values of z, and will therefore serve as a formula of interpolation for fractional or imaginary ones.

The cases where we can find expressions for $f^z(x)$, when $f(x)$ is irrational or above the first degree, are very rare, but as they are (as we have seen) connected so intimately with equations of differences, we will present one or two more in this place, proposing to make these equations the subject of a subsequent Memoir, for which we shall reserve the remainder of our examples.

If $f(x) = \dfrac{2x}{1 - x^2}$, we have $u^2_z \cdot u_{z+1} - u_{z+1} + 2u_z = 0$

an equation whose complete integral is

$$u_z = \sqrt{(-1)} \cdot \frac{1 - c^{2^z}}{1 + c^{2^z}} ;$$

c being an arbitrary constant. Thus,

$$u_0 = x = \sqrt{(-1)} \cdot \frac{1 - c}{1 + c},$$

from which equation, we find

$$c = \frac{1 + x \cdot \sqrt{(-1)}}{1 - x \cdot \sqrt{(-1)}}$$

consequently,

$$f^z(x) = \frac{1}{\sqrt{(-1)}} \left\{ \frac{\{1 + x \cdot \sqrt{(-1)}\}^{2^z} - \{1 - x \cdot \sqrt{(-1)}\}^{2^z}}{\{1 + x \cdot \sqrt{(-1)}\}^{2^z} + \{1 - x \cdot \sqrt{(-1)}\}^{2^z}} \right\}$$

If $f(x) = 2x^2 - 1$, we have $\quad 0 = u_{z+1} - 2 \cdot u^2_z + 1$

and, integrating, $\qquad\qquad u_z = \frac{1}{2}(c^z + c^{-z})$

c being an arbitrary constant.

If we make $u_0 = x$, as before, we find $0 = 1 - 2cx + c^2$, whence

$$f^z(x) = \frac{1}{2}\left\{ \{x + \sqrt{(x^2-1)}\}^z + \{x - \sqrt{(x^2-1)}\}^z \right\}$$

Note II.

The sign S is made use of in preference to Σ, not only to avoid the complication attendant on expressing the $(i+1)^{th}$ term of the series, in place of the ith; but, which is of more importance, to avoid introducing the symbols and algorithm of the integral calculus, in the course of investigations from which the ideas annexed to them, and the operations of that calculus are excluded,——— the shadow of deep research without the substance. The best Authors have not always been scrupulously careful in this point, the varying their notation with their subject, and keeping the former subordinate to the latter. Thus in the *Calcul des Fonctions*, the system of accentuation, however well adapted to explain the general theory of functions, becomes unmanageable when applied to particular cases, or functions actually expressed, and inappropriate, when function in general is no longer the prominent idea. Yet to this system the subject is forced to accommodate itself, to the no small embarrassment of the reader who is not already familiar with it by other roads.

Note III.

The Equation $\{1,1\}$ in the form it here stands, involving no differential process, is free from any difficulty which may arise from supposing n fractional, negative or imaginary. " It thus," to use the phrase of a celebrated Author*, " acquires a greater generality, and supposes no longer that n is an

---

* Legendre; Exercises de calcul Integral, page 279.

integer," or even real. Let then $n = \dfrac{\phi}{\theta \sqrt{(-1)}}$, transferring our ideas from numbers to mere symbols; and we have

$$S\left\{\varepsilon^{i\phi}.f\{x.\varepsilon^{i\theta\sqrt{(-1)}}\} + \varepsilon^{-i\phi}.f\{x.\varepsilon^{-i\theta\sqrt{(-1)}}\}\right\} = -f(x)$$

Let the nth differential coefficient be taken with respect to ϕ, and we get

$$0 = S\left\{i^n.\left(f\{x.\varepsilon^{i\theta\sqrt{(-1)}}\}.\varepsilon^{i\phi} + (-1)^n.f\{x.\varepsilon^{-i\theta\sqrt{(-1)}}\}.\varepsilon^{-i\phi}\right)\right\}$$

But the operation by which we have arrived at this conclusion has been so absolutely singular, that we cannot leave the subject thus abruptly. We have already exhibited algebraic expressions for certain z^{th} functions in terms of z; such as,

$$f(x) = \frac{ax}{b+cx}; \quad \text{and} \quad f^z(x) = \frac{a^z.x}{b^z + cx.\left(\frac{a^z - b^z}{a-b}\right)}$$

When z is a positive or negative integer, this gives us the value of $f^z(x)$ in a form capable of being verified by trial. But, if z be fractional or imaginary, the only meaning we can assign to $f^z(x)$ is, that *it is* that function of z and x which is here connected to it by the sign of equality. In this sense, and way of conceiving it, we may take the differential coefficient with respect to z, as well as x. Thus in the case before us,

$$\left(\frac{df^z(x)}{dz}\right) = x.\left(1 - \frac{cx}{a-b}\right).\left(\frac{a}{b}\right)^z.\log.\left(\frac{a}{b}\right).\left\{1 - \left(\frac{cx}{a-b}\right).\left(1 - \left(\frac{a}{b}\right)^z\right)\right\}^{-2}$$

In the same way, if we have

$$\mathbf{D}^n f(x) = \phi(n, x)$$

When n is not an integer, we may look upon this equation as the definition of $\mathbf{D}^n f(x)$, and thus n is reduced to a mere symbol. In this view of the subject we shall offer one or two singular, and not inelegant formulæ, depending on imaginary, and fractional differentiation.

We know, that $\dfrac{d^n a^x}{dx^n} = a^x.(\log.a)^n$

For n write successively $\theta\sqrt{(-1)}$, and $-\theta\sqrt{(-1)}$, and we obtain, supposing $a = \varepsilon^{\epsilon}$

$$\frac{1}{2} \left\{ \left(\frac{d^{\theta \sqrt{(-1)}} . \varepsilon^{\epsilon x}}{dx^{\theta \sqrt{(-1)}}} \right) + \left(\frac{d^{-\theta \sqrt{(-1)}} . \varepsilon^{\epsilon x}}{dx^{-\theta \sqrt{(-1)}}} \right) \right\} = \varepsilon^{\epsilon x}. \cos. \theta$$

and

$$\frac{1}{2\sqrt{(-1)}} \left\{ \left(\frac{d^{\theta \sqrt{(-1)}} . \varepsilon^{\epsilon x}}{dx^{\theta \sqrt{(-1)}}} \right) - \left(\frac{d^{-\theta \sqrt{(-1)}} . \varepsilon^{\epsilon x}}{dx^{-\theta \sqrt{(-1)}}} \right) \right\} = \varepsilon^{\epsilon x}. \sin. \theta$$

and consequently,

$$\frac{\left(\frac{d^{\theta \sqrt{(-1)}} . \varepsilon^{\epsilon x}}{dx^{\theta \sqrt{(-1)}}} \right) - \left(\frac{d^{-\theta \sqrt{(-1)}} . \varepsilon^{\epsilon x}}{dx^{-\theta \sqrt{(-1)}}} \right)}{\left(\frac{d^{\theta \sqrt{(-1)}} . \varepsilon^{\epsilon x}}{dx^{\theta \sqrt{(-1)}}} \right) + \left(\frac{d^{-\theta \sqrt{(-1)}} . \varepsilon^{\epsilon x}}{dx^{-\theta \sqrt{(-1)}}} \right)} = \sqrt{(-1)} . \tan. \theta$$

Let us next consider the form of mixed differences,

$$\left(\frac{d^m \Delta^n u_x}{dn^m} \right)$$

we know that

$$\Delta^n u_x = u_{x_n} - \frac{n}{1} . u_{x_{n-1}} + \frac{n(n-1)}{1.2} . u_{x_{n-2}} - \&c. = \Delta^n u_{x_0}$$

consequently,

$$\left(\frac{d \Delta^n u_{x_0}}{dn} \right) = \left(\frac{d u_{x_n}}{dn} \right) - \frac{n}{1} . \left(\frac{d u_{x_{n-1}}}{d(n-1)} \right) + \frac{n(n-1)}{1.2} . \left(\frac{d u_{x_{n-2}}}{d(n-2)} \right) - \&c.$$

$$- \frac{Dn}{1} . u_{x_{n-1}} + \frac{D\{n.(n-1)\}}{1.2} . u_{x_{n-2}} - \&c.$$

$$= \Delta^n \left(\frac{d u_{x_0}}{do} \right) + S \left\{ (-1)^i . \frac{D\{n.(n-1)....(n-i+1)\}}{1.2.....i} . u_{x_{n-i}} \right\}$$

Now,

$$\left(\frac{d u_{x_z}}{dz} \right) = \left(\frac{d u_{x_z}}{dx_z} \right) . \left(\frac{d x_z}{dz} \right) = \left(\frac{d u_{x_z}}{dx_z} \right) . \Delta x_0$$

for $x_z = x_0 + z . \Delta x_0$; and consequently,

$$\left(\frac{d u_{x_0}}{do} \right) = \left(\frac{d u_{x_0}}{d x_0} \right) . \Delta x_0 = \left(\frac{d u_x}{d x} \right) . \Delta x$$

o

Thus we have,

$$\left(\frac{d\,\Delta^n u_x}{dn}\right) = \Delta x \cdot \Delta^n \left(\frac{du_x}{dx}\right) + S\left\{(-1)^i \cdot \frac{D\{n.(n-1)....(n-i+1)\}}{1.2.3....i} \cdot u_{x_{n-i}}\right\}..(a)$$

Now for u_x, write $\left(\dfrac{du_x}{dx}\right)$, and the equation becomes

$$\left\{\frac{d\,\Delta^n \left(\frac{du_x}{dx}\right)}{dn}\right\} =$$

$$= \Delta x \cdot \Delta^n \left(\frac{d^2 u_x}{dx^2}\right) + S\left\{(-1)^i \cdot \frac{D\{n(n-1)....(n-i+1)\}}{1.2....i} \cdot \left(\frac{du_{x_{n-i}}}{dx_{n-i}}\right)\right\}$$

Take then the differential coefficients of (a) with respect to n, and for $\left\{\dfrac{d\,\Delta^n \left(\frac{du_x}{dx}\right)}{dn}\right\}$ in the second member, write this its value; and we find

$$\left.\begin{aligned}
\left(\frac{d^2\,\Delta^n u_x}{dn^2}\right) &= (\Delta x)^2 \Delta^n \left(\frac{d^2 u_x}{dx^2}\right) + \\[2mm]
&+ \frac{2}{1} \cdot S\left\{(-1)^i \cdot \frac{D\{n(n-1)....(n-i+1)\}}{1.2....i} \cdot \left(\frac{du_{x_{n-i}}}{dx_{n-i}}\right)\right\} \cdot \Delta x \\[2mm]
&+ \frac{2.1}{1.2} \cdot S\left\{(-1)^i \cdot \frac{D^2\{n(n-1)....(n-i+1)\}}{1.2....i} \cdot u_{x_{n-i}}\right\}
\end{aligned}\right\} ....(b)$$

In the equation (a), write instead of u_x, $\left(\dfrac{d^2 u_x}{dx^2}\right)$, and we find

$$\left\{\frac{d\,\Delta^n \left(\frac{d^2 u_x}{dx^2}\right)}{dn}\right\} =$$

$$= \Delta x \cdot \Delta^n \left(\frac{d^3 u_x}{dx^3}\right) + S\left\{(-1)^i \cdot \frac{D\{n(n-1)....(n-i+1)\}}{1.2....i} \cdot \left(\frac{d^2 u_{x_{n-i}}}{dx^2_{n-i}}\right)\right\}$$

which being substituted in the differential of (b) with respect to n, gives

$$\left(\frac{d^3\,\Delta^n u_x}{dn^3}\right) = (\Delta x)^3 \cdot \Delta^n \left(\frac{d^3 u_x}{dx^3}\right) +$$

$$+ \frac{3}{1} \cdot S\left\{(-1)^i \cdot \frac{D\{n(n-1)....(n-i+1)\}}{1.2....i} \cdot \left(\frac{d^2 u_{x_{n-i}}}{dx^2_{n-i}}\right)\right\} \cdot (\Delta x)^2$$

$$+ \frac{3.2}{1.2} \cdot S\left\{(-1)^i \cdot \frac{D^2\{n(n-1)\ldots(n-i+1)\}}{1.2\ldots i} \cdot \left(\frac{du_{x_{n-i}}}{dx_{n-i}}\right)\right\} \cdot \Delta x$$

$$+ \frac{3.2.1}{1.2.3} \cdot S\left\{(-1)^i \cdot \frac{D^3\{n(n-1)\ldots(n-i+1)\}}{1.2\ldots i} \cdot u_{x_{n-i}}\right\}$$

Thus, continuing the same course of operations, we arrive at length at the following theorem,

$$\left(\frac{d^m \Delta^n u_x}{dn^m}\right) = (\Delta x)^m \cdot \Delta^n \left(\frac{d^m u_x}{dx^m}\right) +$$

$$+ \frac{m}{1} \cdot S\left\{(-1)^i \cdot \frac{D\{n(n-1)\ldots(n-i+1)\}}{1.2\ldots i} \left(\frac{d^{m-1}u_{x_{n-i}}}{dx_{n-i}^{m-1}}\right)\right\} \cdot (\Delta x)^{m-1}$$

$$+ \frac{m(m-1)}{1.2} \cdot S\left\{(-1)^i \cdot \frac{D^2\{n(n-1)\ldots(n-i+1)\}}{1.2\ldots i} \left(\frac{d^{m-2}u_{x_{n-i}}}{dx_{n-i}^{m-2}}\right)\right\} \cdot (\Delta x)^{m-2} \ldots +$$

$$\ldots + \frac{m\ldots 2.1}{1.2\ldots m} \cdot S\left\{(-1)^i \cdot \frac{D^m\{n(n-1)\ldots(n-i+1)\}}{1.2\ldots i} \cdot u_{x_{n-i}}\right\}$$

We may be led to bestow more attention than has been hitherto done, upon this subject, when we consider, that in many equations, the index of differentiation is actually variable. Suppose, for instance, it were required to interpolate a function $u_{x,y}$ for fractional values of x; $u_{x,y}$ being given by an equation of mixed differences.

$$0 = \Delta^n u_{x,y} + a \Delta^{n-1}\left(\frac{du_{x,y}}{dy}\right) + b \Delta^{n-2}\left(\frac{d^2 u_{x,y}}{dy^2}\right) + \&c.$$

x varying according to the characteristic Δ, and y, according to d. The complete integral of this equation, or the general expression of $u_{x,y}$, is

$$u_{x,y} = \varepsilon^{\frac{y}{a_{(1)}}} \cdot (-a_{(1)})^x \cdot \left(\frac{d^x \phi_1(y)}{dy^x}\right) + \varepsilon^{\frac{y}{a_{(2)}}} \cdot (-a_{(2)})^x \cdot \left(\frac{d^x \phi_2(y)}{dy^x}\right) + \ldots \varepsilon^{\frac{y}{a_{(n)}}} \cdot (-a_{(n)})^x \cdot \left(\frac{d^x \phi_n(y)}{dy^x}\right)$$

$\phi_1(y), \ldots \phi_n(y)$ denoting n arbitrary functions of y, and $a_{(1)}, a_{(2)}, \ldots a_{(n)}$ the n roots of

$$0 = z^n - a z^{n-1} + u z^{n-2} - \&c.$$

This expression then will assist us nothing, unless we have means of assigning the values of $\left(\frac{d^x \phi_1(y)}{dy^x}\right)$ for fractional values of x. To fix our ideas let us take $n = 1$, $\phi_1(y) = \varepsilon^{ky}$, and we have

$$0 = u_{x+1, y} - u_{x, y} + a \cdot \left(\frac{d\, u_{x, y}}{dy} \right)$$

$$u_{x, y} = \varepsilon^{\; y \left(k + \frac{1}{a} \right) + x \,.\, \log .\, (-a k)}$$

since $d^x \phi_1 (y) = \varepsilon^{ky} \cdot k^x$

Now for x substitute $x + \dfrac{p}{q} + r \sqrt{(-1)}$, and we find

$$u_{x, y} = \varepsilon^{\frac{p}{q} + r\sqrt{(-1)}} \; \varepsilon^{\; y \left(k + \frac{1}{a} \right) + x \,.\, \log .\, (-a k)}$$

which it is easy to see, satisfies the condition proposed.

———◆———

Note IV.

BY $(Q + R)_x$, $(K - Q - R)_x$, &c. are not to be understood the products of $Q + R$, &c. with x, but simply the quantities $Q_x + R_x$, $K_x - Q_x - R_x$, &c. In the same way, if we had any combination of operations such as $\dfrac{A_x + B_x}{C_x + D_x}$, to be performed on x, we might express it by the same combination of the characteristics, with the symbol x annexed thus $\left\{ \dfrac{A + B}{C + D} \right\}_x$

———◆———

Note V.

WE propose in this note to give a connected theory of the equations of which $\{12, 1\}$ is an insulated case. That equation as well as the theorem $\{1, 6\}$ has something so very singular in it, as may well demand a more particular attention. Let us then suppose

$$F (x) = A_0 + A_1 \cdot x + A_2 \cdot x^2 + \&c.$$
$$f (x) = a_0 + a_1 \cdot x + a_2 \cdot x^2 + \&c.$$

We form then the following series of equations :

$$A_0 \cdot f (v) = A_0\, a_0 + A_0\, a_1\, v + A_0\, a_2\, v^2 + \&c.$$
$$A_1 \cdot z \cdot f (vx) = A_1\, z \cdot a_0 + A_1\, z\, a_1\, v\, x + A_1\, z\, a_2\, v^2\, x^2 + \&c.$$
$$A_2 \cdot z^2 \cdot f (vx^2) = A_2\, z^2\, a_0 + A_2\, z^2\, a_1\, vx^2 + A_2\, z^2\, a_2\, v^2\, x^4 + \&c.$$
$$\&c. \quad = \quad \&c. \qquad \&c. \qquad \&c. \qquad \&c.$$

and adding vertically, we find

$$S\left\{A_i\, z^i .\, f\left(v\, x^i\right)\right\} = S\left\{a_i\, v^i .\, F\left(z\, x^i\right)\right\}; \quad \begin{bmatrix} i = -1 \\ i = \infty \end{bmatrix}$$

Let us suppose

$$F\left(x\right) = -\log.\,(1-x) = \frac{x}{1} + \frac{x^2}{2} + \&c.$$

and the equation becomes

$$P\left\{(1-z\,x^i)^{a_i v^i}\right\} = P\left\{\varepsilon^{-\frac{z^i}{i} .\, f(v\, x^i)}\right\}\ldots\ldots\ldots \{A\}$$

For x write successively $x.\,\varepsilon^{\theta\sqrt{(-1)}}$, and $x.\,\varepsilon^{-\theta\sqrt{(-1)}}$, and first, multiplying the results together, we obtain

$$P\left\{\left(1 - 2\,z.x^i.\cos.\,i\theta + z^2\,x^{2i}\right)^{a_i v^i}\right\} =$$

$$= P\left\{\varepsilon^{-\frac{z^i}{i}\left(f\{v\,x^i.\,\epsilon^{i\theta\sqrt{(-1)}}\}+f\{v\,x^i.\epsilon^{-i\theta\sqrt{(-1)}}\}\right)}\ldots\ldots\ldots\ldots\ldots\ldots\ldots\ldots\ldots\ldots\right\}$$

Again, let the first result be divided by the second, and the whole raised to the power $\dfrac{1}{2\sqrt{(-1)}}$, and we find

$$P\left\{\varepsilon^{-\frac{z^i}{i}\left\{\frac{f\{v\,x^i.\,\epsilon^{i\theta\sqrt{(-1)}}\} - f\{v\,x^i.\,\epsilon^{-i\theta\sqrt{(-1)}}\}}{2\sqrt{(-1)}}\right\}}\ldots\ldots\ldots\ldots\ldots\ldots\ldots\ldots\ldots\ldots\right\} =$$

$$= P\left\{\left(\frac{1 - z\,x^i.\,\varepsilon^{i\theta\sqrt{(-1)}}}{1 - z\,x^i.\,\varepsilon^{-i\theta\sqrt{(-1)}}}\right)^{\frac{a_i v^i}{2\sqrt{(-1)}}}\right\}$$

$$= P\left\{\varepsilon^{\frac{a_i v^i}{2\sqrt{(-1)}}.\,\log.\left\{\frac{1 - \frac{z\,x^i.\,\sin.\,i\theta}{1-z\,x^i.\,\cos.\,i\theta}\cdot\sqrt{(-1)}}{1 + \frac{z\,x^i.\,\sin.\,i\theta}{1-z\,x^i.\,\cos.\,i\theta}\cdot\sqrt{(-1)}}\right\}}\ldots\ldots\ldots\ldots\ldots\ldots\ldots\ldots\right\} =$$

$$= P\left\{\varepsilon^{-a_i v^i.\tan.^{-1}\left\{\frac{z\,x^i.\,\sin.\,i\theta}{1-z\,x^i.\,\cos.\,i\theta}\right\}}\ldots\ldots\ldots\ldots\ldots\ldots\right\}$$

P

and consequently,

$$S\left\{\frac{z^i}{i}\cdot\frac{f\{vx^i.\varepsilon^{i\theta\sqrt{(-1)}}\}-f\{vx^i.\varepsilon^{-i\theta\sqrt{(-1)}}\}}{2\sqrt{(-1)}}\right\}=S\left\{a_iv^i.\tan.^{-1}\left\{\frac{zx^i\ \sin.i\theta}{1-zx^i.\cos.i\theta}\right\}\right\}$$

Suppose now $z=1$, $x=1$, and, writing 2θ for θ, we get,

$$S\left\{\frac{1}{i}\cdot\frac{f\{v.\varepsilon^{2i\theta\sqrt{(-1)}}\}-f\{v.\varepsilon^{-2i\theta\sqrt{(-1)}}\}}{2\sqrt{(-1)}}\right\}=S\left\{a_iv^i.\tan.^{-1}(\cot an.i\theta)\right\}$$

$$=S\left\{a_iv^i.\left(\frac{2n+1}{2}\pi-i\theta\right)\right\}$$

n being any integer.

Now, $S\left\{a_iv^i.\dfrac{2n+1}{2}\pi\right\}=\dfrac{(2n+1)\pi}{2}.f(v);\qquad S\{a_iv^i.i\theta\}=\theta v.\dfrac{df(v)}{dv}$

Thus we have, at length

$$S\left\{\frac{1}{i}\left(\frac{f\{v\,\varepsilon^{2i\theta\sqrt{(-1)}}\}-f\{v\,\varepsilon^{-2i\theta\sqrt{(-1)}}\}}{2\sqrt{(-1)}}\right)\right\}=\frac{(2n+1)\pi}{2}.f(v)-\theta v.\frac{df(v)}{dv}\ldots\ldots\{B\}$$

Had we supposed $z=-1$, $x=1$, we should have got

$$S\left\{\frac{(-1)^{i+1}}{i}\left(\frac{f\{v\,\varepsilon^{2i\theta\sqrt{(-1)}}\}-f\{v\,\varepsilon^{-2i\theta\sqrt{(-1)}}\}}{2\sqrt{(-1)}}\right)\right\}=n\pi.f(v)+\theta v.\frac{df(v)}{dv}\ldots\ldots\{C\}$$

Let these be added together, and we get {since the n's may be *unequal*}

$$S\left\{\frac{1}{2i-1}\left(\frac{f\{v\,\varepsilon^{(4i-2)\theta\sqrt{(-1)}}\}-f\{v\,\varepsilon^{-(4i-2)\theta\sqrt{(-1)}}\}}{2\sqrt{(-1)}}\right)\right\}=\frac{(2n+1)\pi}{4}.f(v)\ldots\ldots\{D\}$$

For $f(v)$, now write $\log.f(v)$, and these equations become

$$P\left\{\left(\frac{f\{v.\varepsilon^{2i\theta\sqrt{(-1)}}\}}{f\{v.\varepsilon^{-2i\theta\sqrt{(-1)}}\}}\right)^{\frac{1}{i\sqrt{(-1)}}}\right\}=\{f(v)\}^{(2n+1)\pi}\cdot\varepsilon^{-\frac{2v\theta}{f(v)}\cdot\frac{df(v)}{dv}};\qquad\ldots\ldots\ldots\{E\}$$

$$P\left\{\left(\frac{f\{v.\varepsilon^{2i\theta\sqrt{(-1)}}\}}{f\{v.\varepsilon^{-2i\theta\sqrt{(-1)}}\}}\right)^{\frac{(-1)^{i+1}}{i\sqrt{(-1)}}}\right\}=\{f(v)\}^{2n\pi}\cdot\varepsilon^{\frac{2v\theta}{f(v)}\cdot\frac{df(v)}{dv}};\qquad\ldots\ldots\ldots\{F\}$$

$$P\left\{\left(\frac{f'\{v.\varepsilon^{(4i-2)\theta\sqrt{(-1)}}\}}{f\{v.\varepsilon^{-(4i-2)\theta\sqrt{(-1)}}\}}\right)^{\frac{1}{(2i-1)\sqrt{(-1)}}}\right\} = \{f(v)\}^{\frac{2n+1}{2}\pi}\ldots\ldots\{G\}$$

In these equations let us write $\varepsilon^{\tan^{-1}.v}$ for $f(v)$, and by the third of the equations in page 35, we find, putting $\frac{\theta}{2}$ for θ,

$$\left\{\frac{1+2v.\sin.\theta+v^2}{1-2v.\sin.\theta+v^2}\right\}^{\frac{1}{1}}\cdot\left\{\frac{1+2v.\sin.2\theta+v^2}{1-2v.\sin.2\theta+v^2}\right\}^{\frac{1}{2}}\cdot\left\{\frac{1+2v.\sin.3\theta+v^2}{1-2v.\sin.3\theta+v^2}\right\}^{\frac{1}{3}}.\&c.=\varepsilon^{(4n+2)\pi.\tan^{-1}.v-\frac{2v\theta}{1+v^2}}$$

$$\left\{\frac{1+2v.\sin.\theta+v^2}{1-2v.\sin.\theta+v^2}\right\}^{\frac{+1}{1}}\cdot\left\{\frac{1+2v.\sin.2\theta+v^2}{1-2v.\sin.2\theta+v^2}\right\}^{\frac{-1}{2}}\cdot\left\{\frac{1+2v.\sin.3\theta+v^2}{1-2v.\sin.3\theta+v^2}\right\}^{\frac{+1}{3}}.\&c.=\varepsilon^{4n\pi.\tan^{-1}.v+\frac{2v\theta}{1+v^2}}$$

$$\left\{\frac{1+2v.\sin.\theta+v^2}{1-2v.\sin.\theta+v^2}\right\}^{\frac{1}{1}}\cdot\left\{\frac{1+2v.\sin.3\theta+v^2}{1-2v.\sin.3\theta+v^2}\right\}^{\frac{1}{3}}.\&c.=\varepsilon^{(2n+1)\pi.\tan^{-1}.v}$$

If $v=1$, these equations become,

$$\left\{\tan.\left(\frac{\pi}{4}+\theta\right)\right\}^{\frac{1}{1}}\cdot\left\{\tan.\left(\frac{\pi}{4}+2\theta\right)\right\}^{\frac{1}{2}}\cdot\left\{\tan.\left(\frac{\pi}{4}+3\theta\right)\right\}^{\frac{1}{3}}.\&c.=\varepsilon^{(2n+1).\frac{\pi^2}{4}-\theta}$$

$$\left\{\tan.\left(\frac{\pi}{4}+\theta\right)\right\}^{\frac{1}{1}}\cdot\left\{\cot.\left(\frac{\pi}{4}+2\theta\right)\right\}^{\frac{1}{2}}\cdot\left\{\tan.\left(\frac{\pi}{4}+3\theta\right)\right\}^{\frac{1}{3}}\cot.\&c.=\varepsilon^{n.\frac{\pi^2}{2}+\theta}$$

$$\left\{\tan.\left(\frac{\pi}{4}+\theta\right)\right\}^{\frac{1}{1}}\cdot\left\{\tan.\left(\frac{\pi}{4}+3\theta\right)\right\}^{\frac{1}{3}}.\&c.=\ldots\ldots=\varepsilon^{(2n+1)\frac{\pi^2}{8}}$$

This last equation, by writing $\frac{\pi}{4}+\theta$, for θ, becomes.

$$(\tan.\theta)^{\frac{1}{1}}.(\cot.3\theta)^{\frac{1}{3}}.(\tan.5\theta)^{\frac{1}{5}}.(\cot.7\theta)^{\frac{1}{7}}.\&c.=\varepsilon^{-(2n+1)\frac{\pi^2}{8}}$$

which is the complete expression of equation $\{12,1\}$. We should have obtained this, had we taken the general value $(2n+1)\pi\sqrt{(-1)}$ for log. (-1) instead of the particular one $\pi\sqrt{(-1)}$.

It is easy to see the vast variety of singular theorems which may be derived from these principles. We shall content ourselves with a very few instances, as we have already extended this Memoir beyond what (in the opinion of some) the importance of the subject may seem to demand.

In the equation $\{A\}$, suppose $a_i = \dfrac{1}{1 \cdot 2 \cdot \ldots \cdot i}$; $f(x) = \varepsilon^x - 1$, then writing a for z and c for v

$$\left\{(1-ax)^{\frac{c}{1}} \cdot \left\{(1-ax^2)^{\frac{c}{2}} \cdot \left\{(1-ax^3)^{\frac{c}{3}} \cdot \left\{\&\text{c. ad inf.}\right\} = \left\{\varepsilon^{1-\epsilon^{cx}}\right\}^{\frac{a}{1}} \cdot \left\{\varepsilon^{1-\epsilon^{cx^2}}\right\}^{\frac{a^2}{2}} \cdot \left\{\varepsilon^{1-\epsilon^{cx^3}}\right\}^{\frac{a^3}{3}} \cdot \&\text{c.}^* \ldots (a)\right.\right.\right.$$

If in $\{A\}$ we suppose $a_i = \dfrac{1}{1 \cdot 2 \cdot \ldots \cdot i^2}$; $f(x) = \displaystyle\int \dfrac{dx}{x}(\varepsilon^x - 1)$, taken from $x = 0$, the equation (writing as before, a, c, for z, v) becomes

$$\left\{(1-ax)^{\frac{c}{1}} \cdot \left\{(1-ax^2)^{\frac{1}{2}\frac{c}{2}} \cdot \left\{(1-ax^3)^{\frac{1}{3}\frac{c}{3}} \cdot \left\{\&\text{c. ad inf.}\right\} = \left\{\varepsilon^{\int \frac{dx}{x}\left(1-\epsilon^{cx}\right)}\right\}^{a} \cdot \left\{\varepsilon^{\int \frac{dx}{x}\left(1-\epsilon^{cx^2}\right)}\right\}^{a^2} \cdot \left\{\varepsilon^{\int \frac{dx}{x}\left(1-\epsilon^{cx^3}\right)}\right\}^{a^3} \&\text{c..}(b)\right.\right.\right.$$

if $x = 1$, this equation becomes *evidently* identical.

Again, suppose $a_i = \dfrac{1}{1 \cdot 2 \cdot 3 \cdot \ldots \cdot i^{n+1}}$; $f(x) = \displaystyle\int^n du^n \cdot (\varepsilon^{\epsilon^u} - 1)$; $\begin{bmatrix} u = -\infty \\ u = \log. x \end{bmatrix}$ and we obtain, as before

$$\left\{(1-ax)^{\frac{c}{1}} \cdot \left\{(1-ax^2)^{\frac{1}{2^n}\frac{c}{2}} \cdot \left\{(1-ax^3)^{\frac{1}{3^n}\frac{c}{3}} \cdot \left\{\&\text{c. ad inf.}\right\} = \left\{\varepsilon^{\int^n du^n\left(1-\epsilon^{c\epsilon^u}\right)}\right\}^{a \cdot 1^{n-1}} \cdot \left\{\varepsilon^{\int^n du^n\left(1-\epsilon^{c\epsilon^{2u}}\right)}\right\}^{a^2 \cdot 2^{n-1}} \&\text{c..}(c)\right.\right.\right.$$

the integrals being taken between the limits $u = -\infty$, and $u = \log. x$.

For n write $-n$, or, suppose

$$a_i = \frac{(1+i)^n}{1 \cdot 2 \cdot 3 \cdot \ldots \cdot i}; \quad f(cx^i) = \frac{1}{i^n \cdot x^{i-1}} \cdot \frac{1}{dx} d \frac{x}{dx} d \frac{x}{dx} d \ldots \frac{x}{dx} d \cdot x^i \left\{\varepsilon^{cx^i} - 1\right\}$$

and we obtain an analogous expression for

$$\left\{(1-ax)^{\frac{c}{1}} \cdot \left\{(1-ax^2)^{2^n \frac{c}{2}} \cdot \left\{(1-ax^3)^{3^n \frac{c}{3}} \cdot \left\{^{4^n}\frac{c}{4}\&\text{c. ad inf.}\right\}\right.\right.\right.$$

---

* A particular case of this, viz. when $a = c = 1$ was communicated to me some time ago in a letter from the Author of the preceding Memoir. I have since, arrived at the equation (a) and (b) by several different and independent methods.

We shall, however, only stop to remark, that the expression for

$$f(x) = 1^n + \frac{2^n x}{1} + \frac{3^n x}{1.2} + \&c. = \frac{1}{dx} d \frac{x}{dx} d \frac{x}{dx} d \ldots \frac{x}{dx} d \cdot x \{\varepsilon^x - 1\}$$

may be exhibited in a very elegant form, without differentiation, and which may be extended to any series of the form

$$\phi(x) = f(0) + \frac{x}{1} \cdot f(1) + \frac{x^2}{1.2} \cdot f(2) + \&c.$$

for,

$$f(i) = f(0) + \frac{i}{1} \cdot \Delta f(0) + \frac{i(i-1)}{1.2} \cdot \Delta^2 f(0) + \&c.$$

Substituting for i successively 0, 1, 2, 3, ... and writing the results in the former equations, we have

$$\phi(x) = f(0) \cdot \left\{ 1 + \frac{x}{1} + \frac{x^2}{1.2} + \&c. \right\}$$

$$+ \frac{\Delta f(0)}{1} \cdot \left\{ \frac{1 \cdot x}{1} + \frac{2 \cdot x^2}{1.2} + \frac{3 \cdot x^3}{1.2.3} + \&c. \right\}$$

$$+ \frac{\Delta^2 f(0)}{1.2} \cdot \left\{ \frac{1.2 \cdot x^2}{1.2} + \frac{2.3 \cdot x^3}{1.2.3} + \&c. \right\}$$

or, simply

$$\phi(x) = \varepsilon^x \left\{ f(0) + \frac{x}{1} \cdot \Delta f(0) + \frac{x^2}{1.2} \cdot \Delta^2 f(0) + \&c. \right\}$$

If $x = 1$, this equation puts on the following remarkable form,

$$f(0) + \frac{f(1)}{1} + \frac{f(2)}{1.2} + \frac{f(3)}{1.2.3} + \&c. = \varepsilon \left\{ f(0) + \frac{\Delta f(0)}{1} + \frac{\Delta^2 f(0)}{1.2} + \&c. \right\}$$

If $f(i)$ be a rational integral function, this terminates, and affords a complete summation of the first member; thus we find

$$1 + \frac{2}{1} + \frac{3}{1.2} + \&c. = 2\varepsilon$$

$$1^2 + \frac{2^2}{1} + \frac{3^2}{1.2} + \&c. = 5\varepsilon$$

$$1^3 + \frac{2^3}{1} + \frac{3^3}{1.2} + \&c. = 15\varepsilon, \quad \&c. \quad \&c. \quad \&c.$$

Q

There exists a curious enough equation of differences between the successive values of the series

$$1^n + \frac{2^n}{1} + \frac{3^n}{1.2} + \&c. = u_n$$

which, since we are engaged on the subject, we will insert

$$u_n = \frac{1^{n+1}}{1} + \frac{2^{n+1}}{1.2} + \&c. = S\left\{\frac{i^{n+1}}{1.2\ldots i}\right\} = S\left\{\frac{\{(i+1)-1\}^{n+1}}{1.2\ldots i}\right\}$$

$$= S\left\{\frac{(i+1)^{n+1}}{1\ldots i}\right\} - \frac{n+1}{1}\cdot S\left\{\frac{(i+1)^n}{1\ldots i}\right\} + \frac{(n+1).n}{1.2}\cdot S\left\{\frac{(i+1)^{n-1}}{1\ldots i}\right\} - \&c.$$

$$= \left\{u_{n+1} - \frac{n+1}{1}\cdot u_n + \frac{(n+1).n}{1.2}u_{n-1} - \&c.\right\} - \left\{1 - \frac{n+1}{1} + \frac{(n+1).n}{1.2} - \&c.\right\}$$

or, $u_n = \Delta^{n+1} u_0$. Since $1 - \frac{n+1}{1} + \frac{(n+1).n}{1.2} - \&c. = (1-1)^{n+1} = 0.$

This gives also

$$u_{n+1} = u_n\left(1 + \frac{n+1}{1}\right) - \frac{(n+1).n}{1.2}u_{n-1} + \frac{(n+1)\ n(n-1)}{1.2.3}u_{n-2} - \&c.$$

from which any value may be found from the preceding.

The sum of the series $u_n = 1^n + \frac{2^n}{1} + \frac{3^n}{1.2} + \&c.$, may be found also by considering that the generating function of $\frac{u_n}{1.2\ldots n}$ is $\varepsilon^{\varepsilon^t}$; for we have

$$\varepsilon^{\varepsilon^t} = 1 + \frac{\varepsilon^t}{1} + \frac{\varepsilon^{2t}}{1.2} + \&c. = \left\{1 + \frac{1}{1} + \frac{1}{1.2} + \&c.\right\} + \frac{t}{1}\left\{1 + \frac{1}{1} + \frac{1}{1.2} + \&c.\right\}$$

$$+ \frac{t^2}{1.2}\left\{1^2 + \frac{2^2}{1} + \frac{3^2}{1.2} + \&c.\right\}$$

$$+ \&c.$$

thus $u_n = 1.2.3\ldots n \times$ the coefficient of t^n in the developement of $\varepsilon^{\varepsilon^t}$,

or,
$$u_n = \frac{d^n.\varepsilon^{\varepsilon^t}}{dt^n}$$

t being supposed $= 0$ after the differentiations.

It remains to deliver some instances of the actual summation of series of tangents. Examples of what we have already given would be uninteresting. We shall therefore have recourse to the theorems (B), (C) of this note. From the expression

$$\tan. A = \frac{1}{\sqrt{(-1)}} \cdot \frac{\varepsilon^{A\sqrt{(-1)}} - \varepsilon^{-A\sqrt{(-1)}}}{\varepsilon^{A\sqrt{(-1)}} + \varepsilon^{-A\sqrt{(-1)}}}$$

it is easy to deduce the following,

$$\tan. A = \frac{1}{\sqrt{(-1)} + \varepsilon^{-A\sqrt{(-1)}}} - \frac{1}{\sqrt{(-1)} + \varepsilon^{A\sqrt{(-1)}}}$$

Let then $f(x) = \dfrac{1}{\sqrt{(-1)} + x^{-1}}$, $f(1) = \dfrac{1 - \sqrt{(-1)}}{2}$, $\dfrac{df(1)}{d1} = \dfrac{-\sqrt{(-1)}}{2}$, $v = 1$, and we find, writing $\frac{1}{2}\theta$ for θ

$$S\left\{\frac{1}{i} \cdot \frac{\tan. i\theta}{+\sqrt{(-1)}}\right\} = \frac{(2n+1)\pi}{2}\left\{1 - \sqrt{(-1)}\right\} + \frac{\theta}{2} \cdot \sqrt{(-1)}$$

But, by a similar process, we must necessarily have found

$$S\left\{\frac{1}{i} \cdot \frac{\tan. i\theta}{-\sqrt{(-1)}}\right\} = \frac{(2n+1)\pi}{2}\left\{1 + \sqrt{(-1)}\right\} - \frac{\theta}{2}\sqrt{(-1)}$$

therefore, subtracting

$$S\left\{\frac{1}{i}\tan. i\theta\right\} = \frac{\sqrt{(-1)}}{2}\left\{-(2n+1)\pi \cdot \sqrt{(-1)} + \theta\sqrt{(-1)}\right\}$$

or,
$$\frac{\tan. \theta}{1} + \frac{\tan. 2\theta}{2} + \frac{\tan. 3\theta}{3} + \&c. = (2n+1)\left(\frac{\pi}{2}\right) - \left(\frac{\theta}{2}\right)$$

and exactly in the same manner, from equation (C),

$$\frac{\tan. \theta}{1} - \frac{\tan. 2\theta}{2} + \frac{\tan. 3\theta}{3} - \&c. = n\pi + \frac{\theta}{2}$$

which added, give

$$\frac{\tan. \theta}{1} + \frac{\tan. 3\theta}{3} + \frac{\tan. 5\theta}{5} + \&c. = \frac{2n+1}{2}\pi.$$

The operation by which these equations have been derived from (B) and (C), is of such a nature, as to leave the mind unsatisfied, and hesitating as to its legitimacy. Such cases are of frequent occurrence in the theory of exponentials; and it must be confessed, that the management of them, so as to avoid drawing conclusions manifestly absurd, is among the most delicate and at the same time interesting points in the whole theory. We seem as it were treading on the very verge of Analysis, on the line which determines truth from falsehood, and feel ourselves placed in the situation of one who fears to pursue to the utmost, the deductions of his reason, through suspicion of some latent error, or mistrust of his own powers.

EQUATIONS OF DIFFERENCES

AND

THEIR APPLICATION TO THE DETERMINATION OF FUNCTIONS FROM GIVEN CONDITIONS.

PART I.

General Theory of Equations of Differences of the first degree, involving one variable only.

IT has long been known to Geometers, that the integration of the equation

$$0 = u_{x+n} + {}^1A_x . u_{x+n-1} + {}^2A_x . u_{x+n-2} + \ldots {}^nA_x . u_x + B_x; \ldots \{1\}$$

may be reduced to the discovery of $n-1$ particular values, which satisfy the same equation deprived of its last term B_x. The methods by which this result has been obtained, have been exceedingly various both in point of principle and execution; and as they have all concurred in leading to the *same* equation deprived of its last term, it might seem reasonable to conclude, that this was the only equation of the form

$$0 = u_{x+n} + {}^1a_x . u_{x+n-1} + \ldots {}^na_x . u_x; \ldots \ldots (a)$$

whose particular integrals possessed the property above mentioned. Were this the case, the operation for the integration of

$$0 = u_{x+n} + {}^1A_x . u_{x+n-1} + \ldots {}^nA_x . u_x; \ldots \ldots \{2\}$$

perpetually recurring into itself, would preclude all hope of any thing farther in the theory of these equations, than what is already known. But happily, this opinion, however apparently well founded, is really erroneous. We shall shew in the course of the present Memoir, that there exists an infinite number of equations of the form (a), essentially different from $\{2\}$, and from each other, which have the same property. We must not then, although no particular integral of {2} should offer itself, conclude on that account, that the equation {1} is unintegrable, since in the employment of other subsidiary equations, we may be more fortunate. It is surprising how this should have escaped the observation of such profound Geometers as have occupied themselves with the subject before us; especially Laplace, who in one of his Memoirs * appears to have considered it in a point of view, remotely similar to that which we have chosen at present; and in one or two cases which will be noticed as they occur, has arrived at the very equations whose consideration, previously to our obtaining a perusal of that admirable Essay, led us to the theory we have now to deliver.

The first observation we have to offer is, that there exists a marked analogy between the constitution of the function

$$u_{x+n} + {}^1A_x . u_{x+n-1} + \dots {}^nA_x . u_x + B_x ; \dots \dots (b)$$

and that of the polynomial

$$u^n + {}^1A . u^{n-1} + \dots \dots {}^nA . u^0,$$

the last term B_x making an exception, as it is not included in the law of the preceding ones.

Let us suppose,

$$(1) \dots \dots u_x^{(1)} = u_{x+1} + {}^1a_x . u_x$$
$$(2) \dots \dots u_x^{(2)} = u_{x+1}^{(1)} + {}^2a_x . u_x^{(1)}$$
$$\vdots$$
$$(n-1) \dots \dots u_x^{(n-1)} = u_{x+1}^{(n-2)} + {}^{n-1}a_x . u_x^{(n-2)}$$
$$(n) \dots \dots u_x^{(n)} = u_{x+1}^{(n-1)} + {}^na_x . u_x^{(n-1)} + B_x$$

$$\Bigg\} ; \dots \dots \{3\}$$

Then, by the successive elimination of the characteristics $u^{(1)}$, $u^{(2)}$, &c. we obtain the following series of equations,

$$u_x^{(1)} = u_{x+1} + {}^1a_x . u_x$$
$$u_x^{(2)} = u_{x+2} + u_{x+1} . \{{}^2a_x + {}^1a_{x+1}\} + u_x . {}^2a_x . {}^1a_x$$
$$u_x^{(3)} = u_{x+3} + u_{x+2} . \{{}^3a_x + {}^2a_{x+1} + {}^1a_{x+2}\} + u_{x+1}\{{}^3a_x . {}^2a_x + {}^3u_x . {}^1a_{x+1} + {}^2a_{x+1} . {}^1a_{x+1}\} + u_x . {}^3a_x . {}^2a_x . {}^1a_x$$

&c. = &c.

$$\vdots$$

---

* Savans Etrangers. 1773.

$$u_x^{(n)} = u_{x+n} + u_{x+n-1} \cdot \left\{ {}^n a_x + {}^{n-1}a_{x+1} + {}^{n-2}a_{x+2} + \ldots {}^1 a_{x+n-1} \right\}$$

$$+ u_{x+n-2} \cdot \begin{cases} {}^n a_x \left\{ {}^{n-1}a_x + {}^{n-2}a_{x+1} + \ldots {}^1 a_{x+n-2} \right\} \\ + {}^{n-1}a_{x+1} \left\{ {}^{n-2}a_{x+1} + \ldots \ldots {}^1 a_{x+n-2} \right\} \\ \;\; \vdots \\ \;\; \vdots \\ + {}^2 a_{x+n-2} \cdot {}^1 a_{x+n-2} \end{cases}$$

$$+ \; \&c.$$
$$\vdots$$
$$+ \; u_x \cdot {}^n a_x \cdot {}^{n-1}a_x \ldots {}^1 a_x$$
$$+ \; B_x$$

Let this be compared with the expression (b), and we obtain

$$\left. \begin{aligned} (1)\ldots\ldots {}^1 A_x &= {}^n a_x + {}^{n-1}a_{x+1} + {}^{n-2}a_{x+2} + \ldots {}^1 a_{x+n-1} \\ (2)\ldots\ldots {}^2 A_x &= {}^n a_x \cdot {}^{n-1}a_{x+1} + {}^n a_x \cdot {}^{n-2}a_{x+2} + {}^{n-1}a_{x+1} \cdot {}^{n-2}a_{x+1} + \&c. \\ (3)\ldots\ldots {}^3 A_x &= {}^n a_x \cdot {}^{n-1}a_x \cdot {}^{n-2}a_x + \&c. \\ &\;\vdots \\ (n)\ldots\ldots {}^n A_x &= {}^n a_x \cdot {}^{n-1}a_x \ldots\ldots {}^1 a_x \end{aligned} \right\} ;\ldots\ldots\{4\}$$

where the analogy alluded to is sufficiently evident. In the case when ${}^1 a_x \ldots$ &c. and of course ${}^1 A_x \ldots$ &c. are constant, this affords a complete integration of the equation,

$$0 = u_{x+n} + {}^1 A \cdot u_{x+n-1} + \ldots\ldots {}^n A \cdot u_x + B_x$$

for the quantities ${}^1 a \ldots$ &c., being given by the equations

$$\left. \begin{aligned} {}^1 A &= {}^1 a + {}^2 a + \ldots {}^n a \\ {}^2 A &= {}^1 a \cdot {}^2 a + {}^1 a \cdot {}^3 a + \&c. \\ &\;\vdots \\ {}^n A &= {}^1 a \cdot {}^2 a \ldots\ldots {}^n a \end{aligned} \right\}$$

are evidently the n roots of the equation

$$0 = u^n - {}^1 A \cdot u^{n-1} + \ldots\ldots \pm {}^n A.$$

Thus the difficulty is reduced to the integration of the equations {3}, which being of the first order, are easily solved, ($u_x^{(n)}$ being $= 0$, by the supposition).—

If $B_x = 0$, and we suppose for convenience's sake, the signs of $^1a \ldots$ &c. changed, so that they become the n roots of

$$0 = u^n + {}^1A \cdot u^{n-1} + \ldots \ldots + {}^nA$$

we have the following equations

$$\left.\begin{array}{l} 0 = u_{x+1}^{(n-1)} - {}^na \cdot u_x^{(n-1)} \\ u_x^{(n-1)} = u_{x+1}^{(n-2)} - {}^{n-1}a \cdot u_x^{(n-2)} \\ \vdots \\ u_x^{(1)} = u_{x+1} - {}^1a \cdot u_x \end{array}\right\} \text{ whence, } \left\{\begin{array}{l} u_x^{(n-1)} = C_1 \cdot {}^na^x \\ u_x^{(n-2)} = C_2 \cdot {}^{n-1}a^x + C_1 \cdot {}^{n-1}a^{x-1} \cdot \Sigma \left(\dfrac{{}^na}{{}^{n-1}a}\right)^x \\ \vdots \\ u_x = C_n \cdot {}^1a^x + C_{n-1} \cdot {}^1a^{x-1} \cdot \Sigma \left(\dfrac{{}^2a}{{}^1a}\right)^x + C_{n-2} \cdot \text{&c.} \end{array}\right.$$

and by performing the integrations, and writing $^1C, {}^2C \ldots {}^nC$, for the several constant coefficients

$$u_x = {}^1C \cdot {}^1a^x + {}^2C \cdot {}^2a^x + \ldots \ldots {}^nC \cdot {}^na_x$$

In general;—the difficulty of integrating $\{1\}$, is reduced to the discovery of $^1a_x, {}^2a_x \ldots$ &c. or of ^1a_x, and such functions of the rest as will suffice for determining $u^{(1)}$, or any other similar combinations. Now it is evident, that since there are n unknown functions $^1a_x \ldots$ &c., and also n equations $\{4\}$, the elimination of all but one of them, gives an equation for determining that one. But here we have an important remark to make,—that as B_x does not enter into these equations, none of the a_x can be functions of B_x. Of course the operations necessary for determining them, must be the same, whatever value we assign to B_x, and consequently the same as if $B_x = 0$. Thus we arrive at once at this theorem, that " *the equation*

$$0 = u_{x+n} + {}^1A_x \cdot u_{x+n-1} + \ldots {}^nA_x \cdot u_x + B_x$$

" *is integrable in the same cases as the equation*

$$0 = u_{x+n} + {}^1A_x \cdot u_{x+n-1} + \ldots {}^nA_x \cdot u_x "$$

The integral of the former depends therefore on that of the latter, and may be derived from it. It will be interesting then to obtain a general formula for this purpose, which being once deduced, we need have no farther regard to the term B_x, but direct all our attention to the integration of $\{2\}$. Now this may be accomplished as follows. It is easily seen, that (by the integration of the equations {3})

$$u_x = {}^1C . P\{-{}^1a_{x-1}\} + {}^2C . P\{-{}^1a_{x-1}\} . \Sigma . \frac{P\{-{}^2a_{x-1}\}}{P\{-{}^1a_x\}} + \dots$$

$$\dots\dots + {}^nC . P\{-{}^1a_{x-1}\} \Sigma . \frac{P\{-{}^2a_{x-1}\}}{P\{-{}^1a_x\}} \Sigma . \dots . \Sigma . \frac{P\{-{}^na_{x-1}\}}{P\{-{}^{n-1}a_x\}} \Bigg\} *; \dots \{5\}$$

$$+ P\{-{}^1a_{x-1}\} \Sigma . \frac{P\{-{}^2a_{x-1}\}}{P\{-{}^1a_x\}} \Sigma . \dots . \Sigma . \frac{P\{-{}^na_{x-1}\}}{P\{-{}^{n-1}a_x\}} \Sigma . \frac{-B_x}{P\{-{}^na_x\}}$$

that is, of the form

$$u_x = {}^1C . {}^{(1)}u_x + {}^2C . {}^{(2)}u_x + \dots {}^nC . {}^{(n)}u_x + E_x ; \dots\dots \{5,1\}$$

Let us now suppose $B_x = 0$, and consequently $E_x = 0$, then we have

$$u_x = {}^1C . {}^{(1)}u_x + {}^2C . {}^{(2)}u_x + \dots {}^nC . {}^{(n)}u_x ; \dots\dots\dots \{5,2\}$$

${}^{(1)}u_x \dots$ &c. being functions of ${}^1a_x \dots$ &c. are independent on B_x, and consequently are the same in this expression and in {5,1}. Moreover, the integral of {2} being obtained, ${}^1a_x \dots$ &c. are determined, and thus the term

$$E_x = P\{-{}^1a_{x-1}\} . \Sigma . \frac{P\{-{}^2a_{x-1}\}}{P\{-{}^1a_x\}} \Sigma . \dots . \Sigma . \frac{-B_x}{P\{-{}^na_x\}}; \dots \{6\}$$

is determined also. All then that remains to be done in order to deduce the integral of {1} from that of {2}, is to find ${}^1a_x \dots$ &c. in terms of ${}^{(1)}u_x \dots$ &c., which are the several particular integrals of this latter, and which we here suppose (by any means) known. The values of ${}^1a_x \dots$ &c. so found, being substituted in {6} will give the term E_x, which must be added to the complete integral of {2}, to give that of {1}. Now this is an operation of no difficulty, and may be performed by merely comparing the two expressions for u_x in the equations {5} and {5,1}. Thus we obtain,

---

* $P\{\phi_x\}$ is used to denote $\phi_x . \phi_{x-1} . \phi_{x-2} \dots . \phi_k$, k being constant. If x be supposed positive, and $k=1$, we have $P\{\phi_x\} = \phi_1 . \phi_2 \dots . \phi_x$. Laplace has used $\nabla\{\phi_x\}$ in the same sense. The full point after the sign Σ, as usual, extends the integration denoted over *all* the rest of the term.

$$
\left.
\begin{aligned}
(1)\ldots\ldots\ & {}^{(1)}u_x = P\{-{}^1a_{x-1}\} \\[2mm]
(2)\ldots\ldots\ & {}^{(2)}u_x = P\{-{}^1a_{x-1}\}\,\Sigma\boldsymbol{.}\,\frac{P\{-{}^2a_{x-1}\}}{P\{-{}^1a_x\}} \\[2mm]
&\ \vdots \\[2mm]
(n)\ldots\ldots\ & {}^{(n)}u_x = P\{-{}^1a_{x-1}\}\,\Sigma\boldsymbol{.}\,\frac{P\{-{}^2a_{x-1}\}}{P\{-{}^1a_x\}}\,\Sigma\boldsymbol{.}\ldots\ldots\Sigma\boldsymbol{.}\,\frac{P\{-{}^na_{x-1}\}}{P\{-{}^{n-1}a_x\}}
\end{aligned}
\right\}\ ;\ldots\ldots\{7\}
$$

from which we derive the following values of ${}^1a_x\ldots$ &c.

$$
\left.
\begin{aligned}
(1)\ldots\ldots\ -{}^1a_x &= \frac{{}^{(1)}u_{x+1}}{{}^{(1)}u_x} \\[3mm]
(2)\ldots\ldots\ -{}^2a_x &= \frac{{}^{(1)}u_{x+2}}{{}^{(1)}u_{x+1}}\cdot\frac{\Delta\left\{\dfrac{{}^{(2)}u_{x+1}}{{}^{(1)}u_{x+1}}\right\}}{\Delta\left\{\dfrac{{}^{(2)}u_x}{{}^{(1)}u_x}\right\}} \\[3mm]
(3)\ldots\ldots\ -{}^3a_x &= \frac{{}^{(1)}u_{x+3}}{{}^{(1)}u_{x+2}}\cdot\frac{\Delta\left\{\dfrac{{}^{(2)}u_{x+2}}{{}^{(1)}u_{x+2}}\right\}}{\Delta\left\{\dfrac{{}^{(2)}u_{x+1}}{{}^{(1)}u_{x+1}}\right\}}\cdot\frac{\Delta\left\{\dfrac{\Delta\left\{\dfrac{{}^{(3)}u_{x+1}}{{}^{(1)}u_{x+1}}\right\}}{\Delta\left\{\dfrac{{}^{(2)}u_{x+1}}{{}^{(1)}u_{x+1}}\right\}}\right\}}{\Delta\left\{\dfrac{\Delta\left\{\dfrac{{}^{(3)}u_x}{{}^{(1)}u_x}\right\}}{\Delta\left\{\dfrac{{}^{(2)}u_x}{{}^{(1)}u_x}\right\}}\right\}}
\end{aligned}
\right\}\ ;\ldots\ldots\ldots\ldots\{8\}
$$

and so on to $-{}^na_x$.

These values of ${}^1a_x\ldots$ &c. being substituted in $\{6\}$, give E_x in terms of ${}^{(1)}u_x\ldots$ &c. The operations being performed, we are conducted to the following theorem; that " *if*

$$
u_x = {}^1C\boldsymbol{.}\,{}^{(1)}u_x + {}^2C\boldsymbol{.}\,{}^{(2)}u_x + \ldots\ldots{}^nC\boldsymbol{.}\,{}^{(n)}u_x
$$

" *be the complete integral of the equation*

$$
0 = u_{x+n} + {}^1A_x\boldsymbol{.}\,u_{x+n-1} + \ldots\ldots{}^nA_x\boldsymbol{.}\,u_x
$$

" *then will that of*

$$
0 = u_{x+n} + {}^1A_x\boldsymbol{.}\,u_{x+n-1} + \ldots\ldots{}^nA_x\boldsymbol{.}\,u_x + B_x
$$

" *be expressed by*

$$u_x = {}^1C \cdot {}^{(1)}u_x + {}^2C \cdot {}^{(2)}u_x + \ldots \ldots {}^nC \cdot {}^{(n)}u_x$$

$$+ {}^{(1)}u_x \cdot \Sigma \cdot \Delta \left\{ \frac{{}^{(2)}u_x}{{}^{(1)}u_x} \right\} \Sigma \cdot \Delta \left\{ \frac{\Delta \left\{ \frac{{}^{(3)}u_x}{{}^{(1)}u_x} \right\}}{\Delta \left\{ \frac{{}^{(2)}u_x}{{}^{(1)}u_x} \right\}} \right\} \Sigma \cdot \Delta \left\{ \frac{\Delta \left\{ \frac{\Delta \left\{ \frac{{}^{(4)}u_x}{{}^{(1)}u_x} \right\}}{\Delta \left\{ \frac{{}^{(2)}u_x}{{}^{(1)}u_x} \right\}} \right\}}{\Delta \left\{ \frac{\Delta \left\{ \frac{{}^{(3)}u_x}{{}^{(1)}u_x} \right\}}{\Delta \left\{ \frac{{}^{(2)}u_x}{{}^{(1)}u_x} \right\}} \right\}} \right\} \Sigma \ldots \quad ; \ldots \ldots \{9\}"$$

$$\ldots \ldots \Sigma \left\{ \frac{- \mathbf{B}_x}{{}^{(1)}u_{x+n} \cdot \Delta \left\{ \frac{{}^{(2)}u_{x+n-1}}{{}^{(1)}u_{x+n-1}} \right\} \cdot \Delta \left\{ \frac{\Delta \left\{ \frac{{}^{(3)}u_{x+n-2}}{{}^{(1)}u_{x+n-2}} \right\}}{\Delta \left\{ \frac{{}^{(2)}u_{x+n-2}}{{}^{(1)}u_{x+n-2}} \right\}} \right\} \cdot \Delta \&c.} \right\}$$

The same value of the term $\mathbf{E}_x$ may also be derived by the following method, more complicated indeed, but well adapted to give a clear idea of the nature of these equations, and an insight into the artifices proper to be used in their discussion; for which reason we shall not hesitate to insert it, although leading to no result but what we have already deduced in a much more direct and compendious manner.

Instead of making $\mathbf{B}_x$ enter into the nth of the equations $\{3\}$, let us introduce a term $- \beta_x^{(1)}$ into the first, so as to have

$$\left. \begin{aligned} &(1) \ldots \ldots u_x^{(1)} = u_{x+1} + {}^1a_x \cdot u_x - \beta_x^{(1)} \\ &(2) \ldots \ldots u_x^{(2)} = u_{x+1}^{(1)} + {}^2a_x \cdot u_x^{(1)} \\ &\quad \vdots \\ &(n) \ldots 0 = u_x^{(n)} = u_{x+1}^{(n-1)} + {}^na_x \cdot u_x^{(n-1)} \end{aligned} \right\} ; \ldots \ldots \ldots \{10\}$$

If we denote by ${}^1A_x^{(1)}, {}^2A_x^{(1)}, \ldots {}^{n-1}A_x^{(1)}$, the values of ${}^1A_x, \ldots {}^{n-1}A_x$, which result from the supposition ${}^1a_x = 0$; —— by ${}^1A_x^{(2)}, \ldots {}^{n-2}A_x^{(2)}$, the respective values of ${}^1A_x^{(1)}, \ldots {}^{n-2}A_x^{(1)}$, which result from supposing ${}^2a_x = 0$; and so on, we shall obtain, by elimination and substitution in the equation $\{10\}$, a series of equations as follows:

$$(1)\ldots 0 = u_{x+n} + {}^{1}A_x . u_{x+n-1} + \ldots {}^{n}A_x . u_x - \{\beta^{(1)}_{x+n-1} + {}^{1}A_x^{(1)} . \beta^{(1)}_{x+n-2} + \ldots {}^{n-1}A_x^{(1)} . \beta^{(1)}_x\}$$

$$(2)\ldots 0 = u^{(1)}_{x+n-1} + {}^{1}A_x^{(1)} . u^{(1)}_{x+n-2} + \ldots {}^{n-1}A_x^{(1)} . u^{(1)}_x$$

$$\vdots$$

$$(n-1)\ldots 0 = u^{(n-2)}_{x+2} + {}^{1}A_x^{(n-2)} . u^{(n-2)}_{x+1} + {}^{2}A_x^{(n-2)} . u^{(n-2)}_x$$

$$(n)\ldots 0 = u^{(n-1)}_{x+1} + {}^{1}A_x^{(n-1)} . u^{(n-1)}_x$$

$$\Big\} ; \ldots \ldots \{11\}$$

The comparison of the first of these with equation $\{1\}$, gives

$$0 = \beta^{(1)}_{x+n-1} + {}^{1}A_x^{(1)} . \beta^{(1)}_{x+n-2} + \ldots {}^{n-1}A_x^{(1)} . \beta^{(1)}_x + B_x ; \ldots \ldots \{12\}$$

which is the same as $\{11\}$ (2) with the term B_x annexed to it. Now the equation $\{10\}$ (1) gives

$$u_x = {}^{1}C . P\{-{}^{1}a_{x-1}\} + P\{-{}^{1}a_{x-1}\} . \Sigma \frac{u^{(1)}_x}{P\{-{}^{1}a_x\}} + P\{-{}^{1}a_{x-1}\} . \Sigma \frac{\beta^{(1)}_x}{P\{-{}^{1}a_x\}} ; \ldots \{13\}$$

The two first terms of this, are evidently the complete integral of $\{2\}$, to which $\{1\}$ reduces itself when $B_x = 0$; consequently, the last term must vanish when $B_x = 0$. Let us now examine this term more closely.

$\beta^{(1)}_x$ is given by the equation $\{12\}$. The complete integral of this, by what we have just shewn, will be of the form

$$\beta^{(1)}_x = {}^{1}C^{(1)} . {}^{(1)}u^{(1)}_x + \ldots {}^{n-1}C^{(1)} . {}^{(n-1)}u^{(1)}_x + E^{(1)}_x ;$$

and in order that this should vanish when $B_x = 0$, since the ${}^{(1)}u^{(1)}_x \ldots$ &c. are particular integrals of $\{11\}$ (2) into which B_x does not enter, and of course cannot be functions of B_x; we must have $C^{(1)} = 0$, &c. thus, by $\{13\}$ we must have

$$\beta^{(1)}_x = P\{-{}^{2}a_{x-1}\} \Sigma . \frac{\beta^{(2)}_x}{P\{-{}^{2}a_x\}}$$

$\beta^{(2)}_x$ being given by the equation

$$0 = \beta^{(2)}_{x+n-2} + {}^{1}A_x^{(2)} . \beta^{(2)}_{x+n-3} + \ldots {}^{n-2}A_x^{(2)} . \beta^{(2)}_x + B_x$$

In the same way, we find

$$\beta^{(2)}_x = P\{-{}^{3}a_{x-1}\} \Sigma . \frac{\beta^{(3)}_x}{P\{-{}^{3}a_x\}}$$

and so on, till we arrive at

$$\beta_x^{(n-2)} = P\{-{}^{n-1}a_{x-1}\}\, \Sigma\, \frac{\beta_x^{(n-1)}}{P\{-{}^{n-1}a_x\}}$$

where $\beta_x^{(n-1)}$ is given by the equation

$$0 = \beta_{x+1}^{(n-1)} + {}^1A_x^{(n-1)}.\beta_x^{(n-1)} + B_x$$

which, since ${}^1A_x^{(n-1)} = {}^n a_x$, gives

$$\beta_x^{(n-1)} = P\{-{}^n a_{x-1}\}\, \Sigma\, \frac{-B_x}{P\{-{}^n a_x\}}$$

Uniting all these results, we find for the value of E_x

$$E_x = P\{-{}^1 a_{x-1}\}\, \Sigma\,.\, \frac{P\{-{}^2 a_{x-1}\}}{P\{-{}^1 a_x\}}\, \Sigma \ldots\ldots \Sigma\,.\, \frac{P\{-{}^n a_{x-1}\}}{P\{-{}^{n-1} a_x\}}\, \Sigma\,.\, \frac{-B_x}{P\{-{}^n a_x\}}$$

no constants being added in the integrations; which is the same result that we have arrived at before.

If we include the arbitrary constants under the integral signs, we shall obtain a very simple expression for u_x, since then, the expression {5} reduces itself to this very form. Thus we have

$$u_x = P\{-{}^1 a_{x-1}\}\, \Sigma\,.\, \frac{P\{-{}^2 a_{x-1}\}}{P\{-{}^1 a_x\}}\, \Sigma \ldots\ldots \Sigma\,.\, \frac{P\{-{}^n a_{x-1}\}}{P\{-{}^{n-1} a_x\}}\, \Sigma\,.\, \frac{-B_x}{P\{-{}^n a_x\}}\, ;\ldots\ldots \{14\}$$

an arbitrary constant being added at every integration. In like manner, by considering the constants as included under the integral signs, we may omit the terms

$${}^1C.{}^{(1)}u_x + {}^2C.{}^{(2)}u_x + \ldots.{}^nC.{}^{(n)}u_x$$

in the equation {9}.

As an instance for the application of the foregoing theory, let us take the equation

$$0 = u_{x+n} + {}^1A.u_{x+n-1} + \ldots\ldots.{}^nA.u_x + B$$

which is one of the most frequent occurrence.

For this purpose, let ${}^1a, {}^2a, \ldots.{}^na$, be the n roots of the equation

$$0 = u^n + {}^1A.u^{n-1} + \ldots.{}^nA$$

T

and we know, that $^1a_x = -\,^1a$, $^2a_x = -\,^2a$, &c.   Thus

$$u_x = \,^1a^{x-1}.\Sigma.\frac{^2a^{x-1}}{^1a^x}\,\Sigma\ldots\ldots\Sigma.\frac{^na^{x-1}}{^{n-1}a^x}\,\Sigma.\frac{-\,\mathbf{B}}{^na^x}$$

that is, after all reductions

$$u_x = \,^1\mathbf{C}.\,^1a^x + \,^2\mathbf{C}.\,^2a^x + \ldots\ldots\,^n\mathbf{C}.\,^na^x - \frac{\mathbf{B}}{(1-\,^1a)(1-\,^2a)\ldots(1-\,^na)}$$

or, since (by the theory of equations)

$$(u-\,^1a)(u-\,^2a)\ldots(u-\,^na) = u^n + \,^1\mathbf{A}.u^{n-1} + \ldots\,^n\mathbf{A},$$

and consequently,

$$(1-\,^1a)(1-\,^2a)\ldots(1-\,^na) = 1 + \,^1\mathbf{A} + \,^2\mathbf{A} + \ldots\,^n\mathbf{A};$$

$$u_x = \,^1\mathbf{C}.\,^1a^x + \,^2\mathbf{C}.\,^2a^x + \ldots\,^n\mathbf{C}.\,^na^x - \frac{\mathbf{B}}{1 + \,^1\mathbf{A} + \,^2\mathbf{A} + \ldots\,^n\mathbf{A}};\ldots\ldots\{15\}$$

Being now freed from the necessity of considering the term $\mathbf{B}_x$, we shall confine ourselves exclusively to the integration of

$$0 = u_{x+n} + \,^1\mathbf{A}_x.u_{x+n-1} + \ldots\,^n\mathbf{A}_x.u_x;\ldots\ldots\ldots\{2\}$$

In this case we have simply

$$\left.\begin{aligned}
(1)&\ldots\ldots\ldots u_x^{(1)} = u_{x+1} + \,^1a_x.u_x\\
(2)&\ldots\ldots\ldots u_x^{(2)} = u_{x+1}^{(1)} + \,^2a_x.u_x^{(1)}\\
(3)&\ldots\ldots\ldots u_x^{(3)} = u_{x+1}^{(2)} + \,^3a_x.u_x^{(2)}\\
&\vdots\\
(n)&\ldots\ldots 0 = u_x^{(n)} = u_{x+1}^{(n-1)} + \,^na_x.u_x^{(n-1)}
\end{aligned}\right\};\ldots\ldots\ldots\{16\}$$

Let us first find the equation which determines ^1a_x; and to avoid the complicated calculations into which any attempt at the elimination of the rest immediately from $\{4\}$ would lead us, let us employ, instead of these quantities, their symmetrical functions $^1\mathbf{A}_x^{(1)},\ldots$ &c., and eliminate these, by means of the equations

$$u_x^{(1)} = u_{x+1} + \,^1a_x.u_x;\ldots\ldots\ldots\ldots\{16\}\,(1)$$

$$0 = u_{x+n-1}^{(1)} + \,^1\mathbf{A}_x^{(1)}.u_{x+n-2}^{(1)} + \ldots,^{n-1}\mathbf{A}_x^{(1)}.u_x^{(1)};\ldots\{11\}\,(2)$$

These give the following equation

$$0 = u_{x+n} + u_{x+n-1}\begin{Bmatrix} {}^1a_{x+n-1} \\ + {}^1A_x^{(1)} \end{Bmatrix} + u_{x+n-2}\begin{Bmatrix} {}^1a_{x+n-2} \cdot {}^1A_x^{(1)} \\ + {}^2A_x^{(1)} \end{Bmatrix} + \ldots u_{x+1}\begin{Bmatrix} {}^1a_{x+1} \cdot {}^{n-2}A_x^{(1)} \\ + {}^{n-1}A_x^{(1)} \end{Bmatrix} + u_x \cdot {}^1a_x \cdot {}^{n-1}A_x^{(1)}$$

whence, by comparison with $\{2\}$, we find

$$\left.\begin{aligned}
(1)\ldots\ldots\ldots {}^1A_x &= {}^1A_x^{(1)} + {}^1a_{x+n-1} \\
(2)\ldots\ldots\ldots {}^2A_x &= {}^2A_x^{(1)} + {}^1a_{x+n-2} \cdot {}^1A_x^{(1)} \\
&\vdots \\
(n-1)\ldots\ldots {}^{n-1}A_x &= {}^{n-1}A_x^{(1)} + {}^1a_{x+1} \cdot {}^{n-2}A_x^{(1)} \\
(n)\ldots\ldots\ldots {}^nA_x &= \ldots\ldots\ldots {}^1a_x \cdot {}^{n-1}A_x^{(1)}
\end{aligned}\right\};\ldots\ldots\ldots\{17\}$$

from which, by elimination,

$$\left.\begin{aligned}
(1)\ldots\ldots 0 &= 1 - \frac{{}^1A_x}{{}^1a_{x+n-1}} + \frac{{}^2A_x}{{}^1a_{x+n-1}\cdot{}^1a_{x+n-2}} - \ldots \pm \frac{{}^nA_x}{{}^1a_{x+n-1}\ldots{}^1a_x} \\
(2)\ldots\ldots {}^1A_x^{(1)} &= \frac{{}^2A_x}{{}^1a_{x+n-2}} - \frac{{}^3A_x}{{}^1a_{x+n-2}\cdot{}^1a_{x+n-3}} + \ldots \mp \frac{{}^nA_x}{{}^1a_{x+n-2}\ldots{}^1a_x} \\
&\vdots \\
(n-1)\ldots {}^{n-2}A_x^{(1)} &= \frac{{}^{n-1}A_x}{{}^1a_{x+1}} - \frac{{}^nA_x}{{}^1a_{x+1}\cdot{}^1a_x} \\
(n)\ldots\ldots {}^{n-1}A_x^{(1)} &= \frac{{}^nA_x}{{}^1a_x}
\end{aligned}\right\};\ldots\ldots\ldots\{18\}$$

The first of these determines the function ${}^1a_x$, and the remaining $(n-1)$ express the values of ${}^1A_x^{(1)}\ldots$&c. in terms of this. Now, let us suppose the value of u_x in the equation $\{16\}$ (1), which corresponds to $u_x^{(1)} = 0$, to be v_x, or,

$$0 = v_{x+1} + {}^1a_x \cdot v_x, \quad \text{and} \quad {}^1a_x = -\frac{v_{x+1}}{v_x}$$

Let this value be substituted for ${}^1a_x$ in $\{18\}$ (1), and we find

$$0 = v_{x+n} + {}^1A_x \cdot v_{x+n-1} + \ldots {}^nA_x \cdot v_x$$

which is the same as the original equation, as it ought evidently to be, from the consideration that $v_x = \mathrm{P}\{-{}^1a_{x-1}\}$, and consequently is no other than what we have denoted by $^{(1)}u_x$. It ought, of course, to satisfy the equation $\{2\}$. Thus it is equally difficult, to obtain a *general* expression for $'a_x$ with u_x. But, let us suppose that a *particular* value can be found, and we shall have the corresponding particular values of $'A_x^{(1)} \ldots$ &c. by $\{18\}$, and of course the complete equation

$$0 = u_{x+n-1}^{(1)} + {}'A_x^{(1)} . u_{x+n-2}^{(1)} + \ldots\ldots {}^{n-1}A_x^{(1)} . u_x^{(1)}.$$

If then we can obtain a particular integral of $\{2\}$, we reduce its integration to that of an equation of the first degree and $n-1^{\text{th}}$ order. In general, if we can obtain m particular integrals $^{(1)}u_x, {}^{(2)}u_x, \ldots .^{(m)}u_x$ of $\{2\}$, we may, by mere elimination, reduce it to the $n-m^{\text{th}}$ order, as follows.

It is easy to see, that the value of $'a_x$ will take the form

$$- \frac{{}^{(1)}u_{x+1} + {}'C . {}^{(2)}u_{x+1} + \ldots .^{m-1}C . {}^{(m)}u_{x+1}}{{}^{(1)}u_x + {}'C . {}^{(2)}u_x + \ldots .^{m-1}C . {}^{(m)}u_x}$$

For $'C, \ldots .^{m-1}C$, write successively the same constants multiplied by 1, by $2, \ldots .$by m; and we obtain m values of $'a_x$ essentially different; and of course, m sets of values for $'A_x^{(1)}, \ldots .^{n-1}A_x^{(1)}$. We have therefore m different equations of the form $\{11\}$ (2), by whose help, eliminating

$$u_{x+n-1}^{(1)}, \quad u_{x+n-2}^{(1)}, \ldots\ldots . u_{x+n-m+1}^{(1)}$$

we obtain an equation of the $(n-m)^{\text{th}}$ order for determining $u_x^{(1)}$. Its coefficients are functions of $(m-1)$ arbitrary constants; its integration introduces $n-m$ more; while that of $\{16\}$ (1) which determines u_x from $u_x^{(1)}$ introduces another. The value of u_x so found, contains therefore $(m-1) + (n-m) + 1 = n$ arbitrary constants, and is consequently complete. If $m = n-1$, the equation of the $(n-m)^{\text{th}}$ or first order, is generally integrable; and thus we arrive at the well known theorem, " that if we can by any means obtain $n-1$ particular integrals of $\{2\}$, we are enabled completely to integrate that equation, and of course the equation $\{1\}$." If we admit the process delivered in the Mecanique Celeste (Liv. X. p. 254.) to be a complete integration of the general equation of the second order

$$0 = u_{x+2} + {}'A_x . u_{x+1} + {}^2A_x . u_x$$

(as we shall shew hereafter, with a slight modification, it undoubtedly is), we may extend this theorem to $n-2$ particular integrals.

The equations $\{18\}$ are, (with the proper attention to the different notations, and different methods of treating the subject,) the same which Laplace has deduced in the Memoir above cited (Probl. II. p. 42, 43.) He seems, however, to have overlooked the simple substitution which identifies the equation $\{18\}$ (1) with $\{2\}$. At all events he has passed it unnoticed, and his demonstration of the theorem we have just enunciated, involves a train of reasoning proportionably longer and more intricate.

Let us next eliminate all but $^n a_x$ from the equations

$$0 = u_{x+1}^{(n-1)} + {}^n a_x \cdot u_{x+1}^{(n-1)}$$

$$u_x^{(n-1)} = u_{x+n-1} + {}^1 A_x^{(1)} \cdot u_{x+n-2} + \ldots {}^{n-1} A_x^{(1)} \cdot u_x ;$$

$^1 A_x^{(1)} \ldots$ &c. *here* denoting the values of $^1 A_x \ldots$ &c. respectively, which result from supposing, not $^1 a_x$, but $^n a_x = 0$.

By a process exactly similar to that by which the equations $\{18\}$ were derived, we find

$$
\left.
\begin{aligned}
(1) \ldots\ldots 0 &= 1 - \frac{^1 A_x}{^n a_x} + \frac{^2 A_{x+1}}{^n a_x \cdot {}^n a_{x+1}} - \ldots\ldots \pm \frac{^n A_{x+n-1}}{^n a_x \ldots {}^n a_{x+n-1}} \\[2mm]
(2) \ldots\ldots {}^1 A_x^{(1)} &= \frac{^2 A_x}{^n a_x} - \frac{^3 A_{x+1}}{^n a_x \cdot {}^n a_{x+1}} + \ldots\ldots \mp \frac{^n A_{x+n-2}}{^n a_x \ldots {}^n a_{x+n-2}} \\[1mm]
&\;\vdots \\[1mm]
(n-1) \ldots {}^{n-2} A_x^{(1)} &= \frac{^{n-1} A_x}{^n a_x} - \frac{^n A_{x+1}}{^n a_x \cdot {}^n a_{x+1}} \\[2mm]
(n) \ldots {}^{n-1} A_x^{(1)} &= \frac{^n A_x}{^n a_x}
\end{aligned}
\right\} ; \ldots\ldots\{19\}
$$

the first of which, by writing $-\dfrac{v_x^{(1)}}{v_{x+1}^{(1)}}$ for $^n a_x$, becomes

$$0 = v_{x+n}^{(1)} + \left\{\frac{^{n-1} A_{x+n-2}}{^n A_{x+n-1}}\right\} v_{x+n-1}^{(1)} + \left\{\frac{^{n-2} A_{x+n-3}}{^n A_{x+n-1}}\right\} v_{x+n-2}^{(1)} + \ldots \left\{\frac{1}{^n A_{x+n-1}}\right\} v_x^{(1)} ; \ldots\ldots\{20\}$$

The same theory may be applied to m particular integrals of this equation, as to those of $\{2\}$, but we must here observe that from $\{20\}$ we may deduce an infinite number of other equations which have the same property; for let us suppose $^1 a_x^{(1)} \ldots$ &c. to be the same functions of the particular integrals

U

$^{(1)}v_x^{(1)}\ldots,\,^{(n)}v_x^{(1)}$ of this equation, that $^1a_x\ldots$ &c. are of $^{(1)}u_x\ldots$ &c. $\{8\}$; and by supposing

$$v_x^{(2)} = \frac{1}{\mathrm{P}\{-^na_{x-1}^{(1)}\}}; \quad \text{or,} \quad {}^na_x^{(1)} = -\frac{v_x^{(2)}}{v_{x+1}^{(2)}}$$

and making the following substitutions,

$$^1\mathrm{A}_{1,x} = \frac{n-1\mathrm{A}_{x+n-2}}{n\mathrm{A}_{x+n-1}}; \qquad ^2\mathrm{A}_{1,x} = \frac{n-2\mathrm{A}_{x+n-3}}{n\mathrm{A}_{x+n-1}};\ldots\,^z\mathrm{A}_{1,x} = \frac{n-z\mathrm{A}_{x+n-z-1}}{n\mathrm{A}_{x+n-1}}$$

$$^1\mathrm{A}_{2,x} = \frac{n-1\mathrm{A}_{1,x+n-2}}{n\mathrm{A}_{1,x+n-1}}; \qquad ^2\mathrm{A}_{2,x} = \frac{n-2\mathrm{A}_{1,x+n-3}}{n\mathrm{A}_{1,x+n-1}};\ldots\,^z\mathrm{A}_{2,x} = \frac{n-z\mathrm{A}_{1,x+n-z-1}}{n\mathrm{A}_{1,x+n-1}}$$

we find a second subsidiary equation

$$0 = v_{x+n}^{(2)} + {}^1\mathrm{A}_{2,x}\cdot v_{x+n-1}^{(2)} + \ldots {}^n\mathrm{A}_{2,x}\cdot v_x^{(2)}$$

and, in general,

$$0 = v_{x+n}^{(y)} + {}^1\mathrm{A}_{y,x}\cdot v_{x+n-1}^{(y)} + \ldots {}^n\mathrm{A}_{y,x}\cdot v_x^{(y)};\ldots\ldots\{21\}$$

$^z\mathrm{A}_{y,x}$ being determined by the following equation of partial differences between x, y, and z,

$$^z\mathrm{A}_{y+1,x} = \frac{n-z\mathrm{A}_{y,x+n-z-1}}{n\mathrm{A}_{y,x+n-1}};\ldots\ldots\ldots\{22\}$$

The equation $\{21\}$ has the property so often mentioned, that if m particular integrals can be discovered, we may reduce the equation $\{2\}$ to the $(n-m^{\text{th}})$ order. It is to be observed, that this equation, whatever value (excepting zero) we assign to y, does not recur into $\{2\}$, as indeed will be rendered evident by the integration of $\{22\}$, an operation which may be thus performed:

By writing $n-z$ for z, $y-1$ for y, and $x+n-z-1$ for x, we find

$$^{n-z}\mathrm{A}_{y,x+n-z-1} = \frac{^z\mathrm{A}_{y-1,x+n-2}}{n\mathrm{A}_{y-1,x+2(n-1)-z}}$$

which, multiplied into $\{22\}$, gives, (writing again $y-1$ for y)

$$^z\mathrm{A}_{y,x} = \frac{^z\mathrm{A}_{y-2,x+n-2}}{n\mathrm{A}_{y-1,x+n-1}\cdot{}^n\mathrm{A}_{y-2,x+2(n-1)-z}};\ldots\ldots(c)$$

Now by making $z = n$ in $\{22\}$, we find

$$^n A_{y,x} = \left\{ ^n A_{y-1,\,x+n-1} \right\}^{-1}$$

whence it is easy to see that

$$^n A_{y,x} = \left(^1 C_{x+(n-1)y} \right)^{(-1)^y}$$

$^1 C$ being an arbitrary, functional characteristic.

This being written for $^n A_{y,x}$ in (c), gives

$$^z A_{y,x} = \, ^z A_{y-2,\,x+n-2} \cdot \left\{ \frac{^1 C_{x+y(n-1)}}{^1 C_{x+y(n-1)-z}} \right\}^{(-1)^y}$$

from which we obtain (by a process analogous to that by which we integrate an equation of common differences of the second order in y, consisting only of two terms)

$$^z A_{y-2,\,x+n-2} = \, ^z A_{y-4,\,x+2(n-2)} \cdot \left\{ \frac{^1 C_{x+(y-1)(n-1)-1}}{^1 C_{x+(y-1)(n-1)-(z+1)}} \right\}^{(-1)^y}$$

$$\&c. = \&c.$$

and consequently,

$$^z A_{y,x} = \left\{ ^2 C_{z,x} + (-1)^y \cdot {}^3 C_{z,x} \right\} \cdot \left\{ \frac{^1 C_{x+y(n-1)} \cdot {}^1 C_{x+(y-1)(n-1)-1} \cdot {}^1 C_{x+(y-2)(n-1)-2} \cdot \&c.}{^1 C_{x+y(n-1)-z} \cdot {}^1 C_{x+(y-1)(n-1)-(z+1)} \cdot {}^1 C_{x+(y-2)(n-1)-(z+2)} \cdot \&c.} \right\}^{(-1)^y} ; \ldots \{23\}$$

$^1 C_x$, $^2 C_{z,x}$, $^3 C_{z,x}$, being three arbitrary functions to be determined by the conditions

$$^z A_{0,x} = \, ^z A_x ; \qquad ^n A_{1,x} = \frac{1}{^n A_{x+n-1}} ; \qquad ^z A_{1,x} = \frac{^{n-z} A_{x+n-z-1}}{^n A_{x+n-1}} .$$

Having obtained $n-1$ particular integrals of any equation of the form $\{2\}$, the nth may be deduced from them immediately without the tedious process of elimination above indicated. Suppose, for instance, we had $^{(1)}u_x$, $^{(2)}u_x$, $\ldots$, $^{(n-1)}u_x$, and wished to find $^{(n)}u_x$. The $(n-1)$ first of the equations $\{8\}$ give us the values of $^1 a_x$, $\ldots$, $^{n-1} a_x$, and the equation

$$^n a_x = \, ^1 A_x - \left\{ ^1 a_{x+n-1} + {}^2 a_{x+n-2} + \ldots + {}^{n-1} a_{x+1} \right\} ; \ldots \ldots \{4\}(1)$$

or, in case of greater convenience,

$$^n a_x = \frac{^n A_x}{^1 a_x \cdot {}^2 a_x \ldots \ldots {}^{n-1} a_x} ; \ldots \ldots \ldots \ldots \ldots \ldots \ldots \ldots \{4\}(n)$$

determines $^{n}a_{x}$ from these. Having then the values of $^{1}a_{x}, \ldots ^{n}a_{x}$, the equation $\{7\}$ (n) gives us at once the value of $^{(n)}u_{x}$. If we would include the result of this in a general expression, which should at once exhibit the complete integral of $\{1\}$, in terms of the $n-1$ particular integrals $^{(1)}u_{x}, \ldots ^{(n-1)}u_{x}$ of $\{2\}$, it will suffice to substitute in $\{9\}$ for $^{(n)}u_{x}$, its value so found. But by this means, the elegant symmetry of that expression will be destroyed, without gaining any advantage in compensation.

The equations $\{4\}$ have shewn us what functions the coefficients $^{1}\mathbf{A}_{x} \ldots$ &c. are of the subsidiary quantities $^{1}a_{x} \ldots$ &c., and the equations $\{8\}$ have enabled us to express these in terms of the particular integrals. We may thus accomplish another, and highly interesting point, without which the theory of equations of differences must be considered as incomplete, and that is to determine the constitution of the coefficients of any equation of the first degree, regarded as functions of the particular integrals. It is true, that the same end may be obtained in a manner apparently more direct by eliminating $^{1}\mathbf{C}, \ldots ^{n}\mathbf{C}$, from the $n+1$ equations

$$\left. \begin{array}{l} u_{x} = {}^{1}\mathbf{C}\cdot{}^{(1)}u_{x} + {}^{2}\mathbf{C}\cdot{}^{(2)}u_{x} + \ldots {}^{n}\mathbf{C}\cdot{}^{(n)}u_{x} \\ u_{x+1} = {}^{1}\mathbf{C}\cdot{}^{(1)}u_{x+1} + {}^{2}\mathbf{C}\cdot{}^{(2)}u_{x+1} + \ldots {}^{n}\mathbf{C}\cdot{}^{(n)}u_{x+1} \\ \vdots \\ u_{x+n} = {}^{1}\mathbf{C}\cdot{}^{(1)}u_{x+n} + {}^{2}\mathbf{C}\cdot{}^{(2)}u_{x+n} + \ldots {}^{n}\mathbf{C}\cdot{}^{(n)}u_{x+n} \end{array} \right\} ; \ldots \ldots \{24\}$$

and after making the proper reductions comparing the resulting equation with $\{2\}$. But this method (even in particular cases) is extremely tedious, and far from conducting to a general formula; for which reason, as well as to preserve uniformity of analysis, we shall prefer that before indicated. By this means we arrive at the following equations, $\{25\}$

$$^{1}\mathbf{A}_{x} = (-1)^{1}\cdot\frac{{}^{(1)}u_{x+n}}{{}^{(1)}u_{x+n-1}}\left(1+\frac{\Delta\left\{\frac{{}^{(2)}u_{x+n-1}}{{}^{(1)}u_{x+n-1}}\right\}}{\Delta\left\{\frac{{}^{(2)}u_{x+n-2}}{{}^{(1)}u_{x+n-2}}\right\}}\right)\left(1+\frac{\Delta\left\{\frac{\Delta\left\{\frac{{}^{(3)}u_{x+n-2}}{{}^{(1)}u_{x+n-2}}\right\}}{\Delta\left\{\frac{{}^{(2)}u_{x+n-2}}{{}^{(1)}u_{x+n-2}}\right\}}\right\}}{\Delta\left\{\frac{\Delta\left\{\frac{{}^{(3)}u_{x+n-3}}{{}^{(1)}u_{x+n-3}}\right\}}{\Delta\left\{\frac{{}^{(2)}u_{x+n-3}}{{}^{(1)}u_{x+n-3}}\right\}}\right\}}\right)\left\{1+\&\text{c.}\right\}; \ldots (1)$$

$$
{}^2A_x = (-1)^2 \cdot \left\{ \frac{{}^{(1)}u_{x+n}}{{}^{(1)}u_{x+n-1}} \cdot \Delta\left\{\frac{{}^{(2)}u_{x+n-1}}{{}^{(1)}u_{x+n-1}}\right\} \, \Delta\left\{\frac{{}^{(2)}u_{x+n-2}}{{}^{(1)}u_{x+n-2}}\right\} \cdot (\&c.\ to\ n\ factors) \cdot \frac{{}^{(1)}u_{x+n-1}}{{}^{(1)}u_{x+n-2}} \left(1 + \Delta\left\{\frac{{}^{(2)}u_{x+n-2}}{{}^{(1)}u_{x+n-2}}\right\} \Big/ \Delta\left\{\frac{{}^{(2)}u_{x+n-3}}{{}^{(1)}u_{x+n-3}}\right\} \left(1 + \&c.\ to\ (n-1)\ factors\right) \right.
$$

$$
+ \frac{{}^{(1)}u_{x+n}}{{}^{(1)}u_{x+n-1}} \cdot \Delta\left\{\frac{{}^{(2)}u_{x+n-1}}{{}^{(1)}u_{x+n-1}}\right\} \times \frac{{}^{(1)}u_{x+n-1}}{{}^{(1)}u_{x+n-2}} \cdot \Delta\left\{\frac{{}^{(2)}u_{x+n-2}}{{}^{(1)}u_{x+n-2}}\right\}
$$

$$
\dots + \&c. \quad\dots\dots\dots\dots\dots\dots(2)
$$

$$
\&c. = \&c.
$$

$$
\dots
$$

$$
{}^nA_x = (-1)^n \cdot \left\{ {}^{(1)}u_{x+n} \cdot \Delta\left\{\frac{{}^{(2)}u_{x+n-1}}{{}^{(1)}u_{x+n-1}}\right\} \cdot \Delta\left\{\frac{{}^{(3)}u_{x+n-2}}{{}^{(1)}u_{x+n-2}}\right\} \Big/ \Delta\left\{\frac{{}^{(2)}u_{x+n-2}}{{}^{(1)}u_{x+n-2}}\right\} \cdot \Delta\ \&c.\ (to\ n\ factors) \right.
$$

$$
\left. {}^{(1)}u_x \cdot \Delta\left\{\frac{{}^{(2)}u_x}{{}^{(1)}u_x}\right\} \cdot \Delta\left\{\frac{{}^{(3)}u_x}{{}^{(1)}u_x}\right\} \Big/ \Delta\left\{\frac{{}^{(2)}u_x}{{}^{(1)}u_x}\right\} \cdot \Delta\ \&c.\ (to\ n\ factors) \right\} \quad\dots\dots(n)
$$

x

These equations, although not in their present form symmetrical in appearance, with respect to the characteristics ${}^{(1)}u$, ${}^{(2)}u$, &c., yet must necessarily be so in reality, as is evident from the consideration of the equations {24}, in which they are symmetrically involved, with the exception of the several arbitrary constants with which they are combined. The manner in which this symmetry is produced by the eliminations. Now these disappear by the eliminations. The manner in which this symmetry is produced by the mutual destruction of superfluous factors in the numerators and denominators of the respective fractions, which compose ^1A_x.....&c. will be much more clearly seen by actually evolving one or two particular cases, than by any general investigation. Let us take as instances, the values of ^1A_x, ^2A_x in the equation

$$0 = u_{x+2} + {}^1A_x . u_{x+1} + {}^2A_x . u_x$$

The formulæ $\{25\}$ give

$$^1A_x = (-1)^1 . \frac{{}^{(1)}u_{x+2}}{{}^{(1)}u_{x+1}} \left(1 + \frac{\Delta\left\{\frac{{}^{(2)}u_{x+1}}{{}^{(1)}u_{x+1}}\right\}}{\Delta\left\{\frac{{}^{(2)}u_x}{{}^{(1)}u_x}\right\}}\right) = - \frac{{}^{(1)}u_{x+2}}{{}^{(1)}u_{x+1}} . \frac{\frac{{}^{(2)}u_{x+2}}{{}^{(1)}u_{x+2}} - \frac{{}^{(2)}u_x}{{}^{(1)}u_x}}{\frac{{}^{(2)}u_{x+1}}{{}^{(1)}u_{x+1}} - \frac{{}^{(2)}u_x}{{}^{(1)}u_x}} = - \frac{{}^{(1)}u_x . {}^{(2)}u_{x+2} - {}^{(2)}u_x . {}^{(1)}u_{x+2}}{{}^{(1)}u_x . {}^{(2)}u_{x+1} - {}^{(2)}u_x . {}^{(1)}u_{x+1}};$$

$$^2A_x = (-1)^2 . \frac{{}^{(1)}u_{x+2}}{{}^{(1)}u_{x+1}} . \frac{\Delta\left\{\frac{{}^{(2)}u_{x+1}}{{}^{(1)}u_{x+1}}\right\}}{\Delta\left\{\frac{{}^{(2)}u_x}{{}^{(1)}u_x}\right\}} \times \frac{{}^{(1)}u_{x+1}}{{}^{(1)}u_x} = \frac{{}^{(1)}u_{x+2}}{{}^{(1)}u_x} . \frac{\frac{{}^{(2)}u_{x+2}}{{}^{(1)}u_{x+2}} - \frac{{}^{(2)}u_{x+1}}{{}^{(1)}u_{x+1}}}{\frac{{}^{(2)}u_{x+1}}{{}^{(1)}u_{x+1}} - \frac{{}^{(2)}u_x}{{}^{(1)}u_x}} = + \frac{{}^{(1)}u_{x+1} . {}^{(2)}u_{x+2} - {}^{(2)}u_{x+1} . {}^{(1)}u_{x+2}}{{}^{(1)}u_x . {}^{(2)}u_{x+1} - {}^{(2)}u_x . {}^{(1)}u_{x+1}};$$

and thus the equation becomes

$$0 = u_{x+2} - u_{x+1} . \left\{\frac{{}^{(1)}u_x . {}^{(2)}u_{x+2} - {}^{(2)}u_x . {}^{(1)}u_{x+2}}{{}^{(1)}u_x . {}^{(2)}u_{x+1} - {}^{(2)}u_x . {}^{(1)}u_{x+1}}\right\} + u_x . \left\{\frac{{}^{(1)}u_{x+1} . {}^{(2)}u_{x+2} - {}^{(2)}u_{x+1} . {}^{(1)}u_{x+2}}{{}^{(1)}u_x . {}^{(2)}u_{x+1} - {}^{(2)}u_x . {}^{(1)}u_{x+1}}\right\}; ..\{26\}$$

We have endeavoured in the preceding pages to include under general formulæ, some of the principal results which have hitherto been obtained in the theory of equations of finite differences of the first degree, as well as to add something to the stock of information already accumulated. The principle employed, has been attended with one considerable advantage; that of leading us at once to an intimate knowledge of the constitution of the coefficients of these equations, and of exhibiting at the first glance, the connection between the equations $\{1\}$ and $\{2\}$, so often mentioned. The application of the same principle to the equation of common differentials of the first degree

$$0 = \frac{d^n u}{dx^n} + {}^1A_x . \frac{d^{n-1} u}{dx^{n-1}} + \dots {}^nA_x . u + B_x, \dots\dots(d)$$

is equally simple, and leads by short and easy steps to the general theory of this equation. Our limits, however, will not allow us at present to enter upon this subject.

The similarity between the two equations $\{1\}$ and (d), and between the methods of treating them, was first remarked by Lagrange in a Memoir published in the Melanges de Turin, Vol. 1. where he applies d' Alembert's method of integrating the latter equation ($^{\scriptscriptstyle 1}A_x, \ldots$&c. being constant) to the corresponding case of the former, and from which he derives the usual theorems respecting recurring series. In the 3^d Vol. of the same work, (1763—5) appeared his two celebrated theorems concerning the general equation (d), which were extended by Laplace in the 4th Vol. to equations of finite differences. Of this Memoir, (as far as it relates to the same subject,) that which we have above referred to * is a copy almost verbatim. Finally, Condorcet, in the Mem. de l' Acad. des Sciences, has verified these results; and Cousin ✝ has shewn how Lagrange's method of treating the equation (d), viz. multiplication by a factor, may be applied with equal success, although, it must be confessed, not with equal simplicity, to {1}.

---

* Savans Etrangers. 1773.

✝ Lecons de Calcul Differentiel, et de Calcul Integral, tom. ii. p. 722. et suiv. published in 1777.

PART II.

ON THE INTEGRATION OF CERTAIN PARTICULAR EQUATIONS OF DIFFERENCES.

THE equation of the first order and degree

$$0 = u_{x+1} + A_x . u_x + B_x$$

may, as is well known, be integrated as follows.

For u_x, substitute $v_x . P \{ -A_{x-1} \}$, and we find

$$0 = (v_{x+1} - v_x) . P \{ -A_x \} + B_x$$

that is,

$$\Delta v_x = \frac{-B_x}{P \{ -A_x \}}$$

and integrating with respect to x

$$v_x = C + \Sigma \frac{-B_x}{P \{ -A_x \}} \; ; \quad \text{and consequently,} \quad u_x = P \{ -A_{x-1} \} . \left\{ C + \Sigma \frac{-B_x}{P \{ -A_x \}} \right\}$$

Let us next consider the equation of the first degree and second order. We have already observed, that a process is delivered in the Mecanique Celeste, [*] equivalent to the integration of

$$0 = u_{x+2} + a_x . u_{x+1} + b_x . u_x ; \dots \dots \dots \dots \{27\}$$

and we shall take this opportunity to follow up that process, so as to obtain from it (in every case) a complete integral. But first, for the sake of simplification, we will transform the equation $\{27\}$ into one of the same form, but with only one variable coefficient. For this purpose, assume $u_x = v_x . P \{ -a_{x-2} \}$, and we obtain, writing $-c_{x+2}$ for $\dfrac{b_x}{a_x . a_{x-1}}$

$$0 = v_{x+2} - v_{x+1} - c_{x+2} . v_x ; \dots \dots \dots \dots \{28\}$$

---

[*] Mecan. Cel. liv. IV. p. 197. And again, liv. X. p. 254.

and consequently,

$$\frac{v_x}{v_{x-1}} = 1 + \frac{c_x}{\left(\frac{v_{x-1}}{v_{x-2}}\right)} = 1 + \frac{c_x}{1 + \frac{c_{x-1}}{\left(\frac{v_{x-2}}{v_{x-3}}\right)}} =$$

$$= 1 + \frac{c_x}{1 + \frac{c_{x-1}}{1 + \frac{c_{x-2}}{1 + \cdots \frac{c_i}{\left(\frac{v_{i-1}}{v_{i-2}}\right)}}}}$$

i being constant.

Let us denote by the symbol $F(c_x)$, the continued fraction

$$1 + \frac{c_x}{1 + \frac{c_{x-1}}{1 + \ldots c_i}}$$

and since we may suppose $\left(\frac{v_{i-1}}{v_{i-2}}\right)$ equal to an arbitrary constant, if we suppose this constant unity, we find $F(c_x)$ for a particular value of $\left(\frac{v_x}{v_{x-1}}\right)$, and of course $P\{F(c_x)\}$ for a particular integral of v_x in the equation {28}. Having obtained this, the complete integral may easily be deduced, and it will afford a good example of the application of the theory of Part I.

Thus we have

$$^{(1)}v_x = P\{F(c_x)\} = P\{-'a_{x-1}\}$$

and of course

$$'a_x = -F(c_{x+1}); \qquad {}^2a_x = \frac{-c_{x+2}}{-F(c_{x+1})} = F(c_{x+2}) - 1$$

whence by the application of {7} (2)

$$^{(2)}v_x = P\{F(c_x)\} \Sigma \frac{P\{1 - F(c_{x+1})\}}{P\{F(c_{x+1})\}} = P\{F(c_x)\} \Sigma P\left\{\frac{-c_{x+1}}{c_{x+1} + F(c_x)}\right\}$$

Y

Thus we have the complete integral of {28},

$$v_x = \mathrm{P}\{\mathrm{F}(c_x)\} \left({}'\mathrm{C} + {}^2\mathrm{C} . \Sigma \mathrm{P} \left\{ \frac{-c_{x+1}}{c_{x+1} + \mathrm{F}(c_x)} \right\} \right); \dots \dots \{29\}$$

which multiplied by $\mathrm{P}\{-a_{x-2}\}$, gives that of $\{27\}$.

From the foregoing operations, it is easy to derive the complete integral of

$$0 = u_{x+2} + a_x . u_{x+1} + b_x . u_x + \mathrm{B}_x ; \dots \dots \dots \{30\}$$

The substitution of $v_x . \mathrm{P}\{-a_{x-2}\}$ for u_x, gives

$$0 = v_{x+2} - v_{x+1} - c_{x+2} . v_x + \frac{\mathrm{B}_x}{\mathrm{P}\{-a_x\}}$$

whence, after all reductions,

$$u_x = \mathrm{P}\{-a_{x-2} . \mathrm{F}(c_x)\} \Sigma . \mathrm{P} \left\{ \frac{-c_{x+1}}{c_{x+1} + \mathrm{F}(c_x)} \right\} \Sigma . -\mathrm{B}_x . \mathrm{P} \left\{ \frac{c_{x+1} + \mathrm{F}(c_x)}{a_x . c_{x+2} . \mathrm{F}(c_x)} \right\} ; \dots \{31\}$$

a constant being added at each integration.

This process, in the way we have here delivered it, answers equally for positive and negative values of x *. It is to be observed, that $\mathrm{F}(c_x)$ is merely an abbreviated symbol for *a finite algebraic fraction*, whose numerator and denominator consist of a complication of terms, depending on the combinations of c_x, c_{x-1}, &c. The integration of the equation $\{28\}$ does not then enable us to *sum* the continued fraction $\mathrm{F}(c_x)$. It requires this summation to be previously known by other means; in the very same manner as the integration of

$$0 = u_{x+1} + a_x . u_x + \mathrm{B}_x$$

instead of enabling us to sum the series

$$\frac{\mathrm{B}_{x-1}}{\mathrm{P}\{-a_{x-1}\}} + \dots \dots \frac{\mathrm{B}_i}{\mathrm{P}\{-a_i\}}$$

$\left(\text{which enters into the final expression under the form } \Sigma \, \frac{-\mathrm{B}_x}{\mathrm{P}\{-a_x\}} \right)$ pre-supposes

---

* " Cette equation ne peut donc commeucer a avoir lieu que lorsque $x = $ &c."— " Ces Equations ne commencent point a exister toutes a la fois." This is not the place to enter upon the metaphysic of equations of differences. We will merely observe, that such phrases as these, which tend to embarrass the reader with *imaginary* restrictions, are something worse than inappropriate.

that we possess the means of performing the operation denoted by Σ. In default of these means, nothing remains but to write the series at full length, and in this respect, the signs F and Σ, are placed under the same circumstances. We will add one more remark. If by any means we can discover a particular integral of the equation

$$0 = u_{x+2} - a_{x+2} \cdot u_{x+1} - b_{x+2} \cdot u_x$$

independently on continued fractions, we may then return upon the difficulty, and assign the value of the continued fraction

$$a_x + \cfrac{b_x}{a_{x-1} + \cfrac{b_{x-1}}{a_{x-2} + \&c.}}$$

in some other form.

For instance, the continued fraction.

$$1 + \cfrac{2}{1 + \cfrac{2}{1 + \ldots \cfrac{2}{1}}}$$

in which the number 2 occurs x times, is represented by

$$\frac{2^{x+2} + (-1)^{x+1}}{2^{x+1} + (-1)^x}$$

In a Memoir which I had the honour of communicating to this Society in the beginning of the present year, I exhibited a method of integrating the equation

$$0 = u_{x+1} \cdot u_x + a \cdot u_{x+1} + b \cdot u_x + c$$

of the second degree, with constant coefficients. A particular case of this had been previously integrated by Laplace in the Journal de l' Ecole Polytechnique *, viz.

$$0 = u_{x+1} \cdot u_x - a \left(u_{x+1} - u_x \right) + 1$$

---

* Memoire sur divers points d' Analyse.

Now if in our formulæ * we make

$$a = -1, \quad \beta = \gamma = a, \quad \delta = 1$$

we shall have

$$\varpi = 0, \quad \lambda = a, \quad \mu = \sqrt{(-1)}, \quad \nu = \frac{a - \sqrt{(-1)}}{a + \sqrt{(-1)}}$$

$$u_x = \sqrt{(-1)} \cdot \frac{1 - \mathrm{K} \cdot \nu^x}{1 + \mathrm{K} \cdot \nu^x}$$

that is, supposing $\mathrm{K} = \nu^c$, c being an arbitrary constant,

$$u_x = \frac{1}{\sqrt{(-1)}} \cdot \frac{\nu^{\frac{c+x}{2}} - \nu^{-\frac{c+x}{2}}}{\nu^{\frac{c+x}{2}} + \nu^{-\frac{c+x}{2}}} = \tan \left\{ \frac{c+x}{2\sqrt{(-1)}} \cdot \log \left(\frac{a - \sqrt{(-1)}}{a + \sqrt{(-1)}} \right) \right\}$$

$$= \tan \left\{ (c+x) \cdot \tan^{-1} \left(\frac{-1}{a} \right) \right\}$$

which coincides exactly with Laplace's result. An equation of the second order, somewhat similar in form,

$$0 = u_{x+2} \cdot u_{x+1} \cdot u_x - a \left(u_{x+2} + u_{x+1} + u_x \right) ; \ldots \ldots \ldots \ldots \{32\}$$

admits of the following integration.

Assume,

$$u_x = \sqrt{(a)} \cdot \tan v_x$$

and by throwing $\{32\}$ into the form

$$u_{x+2} = - \frac{a \left(u_{x+1} + u_x \right)}{a - u_{x+1} \cdot u_x}$$

we obtain

$$\tan \left(v_{x+2} \right) = - \frac{\tan v_{x+1} + \tan v_x}{1 - \tan v_{x+1} \cdot \tan v_x} = - \tan \left(v_{x+1} + v_x \right)$$

or,

$$0 = v_{x+2} + v_{x+1} + v_x$$

---

* These formulæ and the Memoir referred to, are to be found in another part of the present Volume.

and integrating

$$v_x = \frac{\alpha - \beta \sqrt{(-1)}}{2}\left(-\frac{1}{2} + \sqrt{(-1)}\cdot\sqrt{\left(\frac{3}{4}\right)}\right)^x + \frac{\alpha + \beta\sqrt{(-1)}}{2}\left(-\frac{1}{2} - \sqrt{(-1)}\sqrt{\left(\frac{3}{4}\right)}\right)^x$$

α and β being two arbitrary constants; that is,

$$v_x = \frac{\alpha - \beta\sqrt{(-1)}}{2}\left(\cos\frac{2\pi}{3} + \sqrt{(-1)}\cdot\sin\frac{2\pi}{3}\right)^x + \frac{\alpha + \beta\sqrt{(-1)}}{2}\left(\cos\frac{2\pi}{3} - \sqrt{(-1)}\cdot\sin\frac{2\pi}{3}\right)^x$$

$$= \alpha\cdot\cos\frac{2\pi x}{3} + \beta\cdot\sin\frac{2\pi x}{3}$$

whence, writing this for v_x, we find

$$u_x = \sqrt{(a)}\cdot\tan\left\{\alpha\cdot\cos\frac{2\pi x}{3} + \beta\cdot\sin\frac{2\pi x}{3}\right\}; \ldots\ldots\ldots \{32,1\}$$

In the same way, the integrals of

$$0 = u_{x+2}\cdot u_{x+1}\cdot u_x + a\,(u_{x+2} - u_{x+1} + u_x); \ldots\ldots\ldots\ldots \{33\}$$

$$0 = u_{x+2}\cdot u_{x+1}\cdot u_x + a\,(-u_{x+2} + u_{x+1} + u_x); \ldots\ldots\ldots \{34\}$$

$$0 = u_{x+2}\cdot u_{x+1}\cdot u_x + a\,(+u_{x+2} + u_{x+1} - u_x); \ldots\ldots\ldots \{35\}$$

may be shewn to be, respectively

$$u_x = \sqrt{(a)}\cdot\tan\left\{\alpha\cdot\cos\frac{\pi x}{3} + \beta\cdot\sin\frac{\pi x}{3}\right\}; \ldots\ldots\ldots\ldots\ldots \{33,1\}$$

$$u_x = \sqrt{(a)}\cdot\tan\left\{\alpha\cdot\left(\frac{1 + \sqrt{(5)}}{2}\right)^x + \beta\cdot\left(\frac{1 - \sqrt{(5)}}{2}\right)^x\right\}; \ldots \{34,1\}$$

$$u_x = \sqrt{(a)}\cdot\tan\left\{\alpha\cdot\left(\frac{-1 + \sqrt{(5)}}{2}\right)^x + \beta\cdot\left(\frac{-1 - \sqrt{(5)}}{2}\right)^x\right\}; \ldots \{35,1\}$$

and upon the same principle, we may integrate many equations of superior orders, such as

$$0 = (u_{x+3}u_{x+2}u_{x+1} + u_{x+3}u_{x+2}u_x + u_{x+3}u_{x+1}u_x + u_{x+2}u_{x+1}u_x) - a\,(u_{x+3} + u_{x+2} + u_{x+1} + u_x)$$

$$0 = (u_{x+3}u_{x+2}u_{x+1} + u_{x+3}u_{x+2}u_x + u_{x+3}u_{x+1}u_x - u_{x+2}u_{x+1}u_x) - a\,(u_{x+3} - u_{x+2} - u_{x+1} - u_x)$$

$$0 = (u_{x+3}u_{x+2}u_{x+1} - u_{x+3}u_{x+2}u_x + u_{x+3}u_{x+1}u_x - u_{x+2}u_{x+1}u_x) + a\,(u_{x+3} - u_{x+2} + u_{x+1} - u_x)$$

&c. = &c.

z

We will next proceed to integrate the equation

$$0 = u_{x+1} u_x + a_x . u_{x+1} + b_x . u_x + c_x ; \dots \dots \dots \{36\}$$

in which a_x, b_x, c_x, denote, as usual, any functions whatever of x.

Let us conceive this to have arisen from the expansion of

$$0 = (u_{x+1} + A_{x+1}) (u_x + A_x) + B_x (u_x + A_x) + C_x$$

which gives, when reduced, and compared with $\{36\}$

$$A_x = a_x ; \qquad B_x = b_x - a_{x+1} ; \qquad C_x = c_x - a_x b_x$$

Assume now, $u_x = \dfrac{v_{x+1}}{v_x} - a_x$, and we find

$$0 = v_{x+2} + (b_x - a_{x+1}) v_{x+1} + (c_x - a_x b_x) . v_x$$

We have already integrated this. Let $^{(1)}v_x$ and $^{(2)}v_x$ represent its two particular integrals, and we shall have

$$u_x = \frac{^{(1)}v_{x+1} + C . ^{(2)}v_{x+1}}{^{(1)}v_x + C . ^{(2)}v_x} - a_x$$

C being an arbitrary constant. In the following cases, the integral may be exhibited without continued fractions,

$$0 = u_{x+1} u_x + (a_x u_{x+1} + a_{x+1} u_x) + c_x$$

or, more generally

$$0 = u_{x+1} u_x + a_x . u_{x+1} + (a_{x+1} + b . \phi_{x+1}) u_x + (a_{x+1} . a_x + b . \phi_{x+1} . a_x + c . \phi_x \phi_{x+1})$$

ϕ_x being any function whatever of x.

The consideration of the different orders of any proposed function, leads, as we have shewn in a previous Memoir, to equations of differences of the first order. We shall here present a few examples of that theory which lead to integrable equations of some little generality, such as in the very imperfect state of our knowledge, respecting equations of a degree superior to the first, cannot but be interesting.

Let us first seek an expression for $f^x(v)$ when $f(v)$ is of the form

$$\frac{v^2 + 2av}{4v - a}$$

Assuming $f^x(v) = u_x$, we find the following equation

$$0 = u_x^2 - 4u_x \cdot u_{x+1} + a \cdot (2u_x + u_{x+1}) ; \dots \dots \dots \dots \{37\}$$

This equation may be integrated, by putting

$$u_x = \frac{a}{2} \cdot \frac{2w_x - 1}{w_x + 1}$$

when (by the mutual destruction of certain terms) it becomes

$$0 = w_{x+1} + 2w_x^2 - 1 ; \dots \dots \dots \dots (e)$$

and integrating,

$$w_x = \frac{1}{2\sqrt{(-1)}} \left\{ c^{\frac{x}{2}} - c^{-\frac{x}{2}} \right\}$$

c being an arbitrary constant.

From this, by making $u_0 = v$, we obtain the value of c,

$$c = \left\{ 1 - \left(\frac{a + 2v}{2a - 2v} \right)^2 \right\}^{\frac{1}{2}} + \left(\frac{a + 2v}{2a - 2v} \right) \cdot \sqrt{(-1)}$$

whence it is easy to see, that *

$$u_x = f^v(v) = \frac{a}{2} \cdot \frac{2 \cdot \sin \left\{ 2^x \cdot \sin^{-1} \left(\frac{a + 2v}{2a - 2v} \right) \right\} - 1}{\sin \left\{ 2^x \cdot \sin^{-1} \left(\frac{a + 2v}{2a - 2v} \right) \right\} + 1}$$

Nearly in the same manner, we obtain an expression for $f^x(v)$, when $f(v)$ is of the form

$$\frac{4av - v^2}{2v + a}$$

This leads to the equation

$$0 = u_x^2 + 2u_x u_{x+1} - a(4u_x - u_{x+1}) ; \dots \dots \dots \{38\}$$

---

* It may be proper to remind the reader, that we continue to use $\sin^{-1}$, $\tan^{-1}$, &c. for what were formerly written arc $(\sin = \dots)$, arc $(\tan = \dots)$ &c. The principles of this notation we have explained elsewhere.

which is reduced to (e), by putting

$$u_x = 2a \cdot \frac{w_x + 1}{2w_x - 1}$$

Thus at length, we find

$$u_x = f^x(v) = 2a \cdot \frac{\sin\left\{2^x \cdot \sin^{-1}\left(\frac{2a+v}{2v-2a}\right)\right\} + 1}{2 \cdot \sin\left\{2^x \cdot \sin^{-1}\left(\frac{2a+v}{2v-2a}\right)\right\} - 1}$$

For another instance, let us suppose

$$f(v) = a + \left\{b - 2ac.v + c.v^2\right\}^{\frac{1}{2}}$$

as before, let $f^x(v) = u_x$, and we get

$$0 = u^2_{x+1} - 2a.u_{x+1} - c.u^2_x + 2ac.u_x + (a^2 - b)$$

Now, any equation of the form

$$0 = u^2_{x+1} + A_x.u^2_x + B_x.u_{x+1} + A_x B_{x-1}.u_x + C_x ; \ldots \ldots \{39\}$$

may be immediately integrated, by assuming

$$0 = w_{x+1} + A_x.w_x + C_x$$

$$0 = u^2_x + B_{x-1}.u_x - w_x$$

By eliminating w, we arrive at the proposed equation, and consequently, its integral will be

$$u_x = -\frac{B_{x-1}}{2} + \left\{\left(\frac{B_{x-1}}{2}\right)^2 + P\left\{-A_{x-1}\right\} \cdot \left(\text{Const.} + \Sigma \frac{-C_x}{P\{-A_x\}}\right)\right\}^{\frac{1}{2}} ; \ldots \ldots \{39, 1\}$$

In the case before us, we find

$$u_x = a + \left\{a^2 + \frac{b - a^2}{1 - c} + C.c^x\right\}^{\frac{1}{2}}$$

C being a function of v, which being determined by making $u_0 = v$, gives

$$C = v^2 - 2av - \frac{b - a^2}{1 - c}$$

and consequently,

$$f^x(v) = a + \left\{ \frac{b - c\,a^2}{1 - c} + c^x\left(v^2 - 2\,a\,v - \frac{b - a^2}{1 - c}\right)\right\}^{\frac{1}{2}}$$

Let $a = 0$, and the general expressions will become

$$f(v) = \{b + c\,v^2\}^{\frac{1}{2}}; \qquad f^x(v) = \left\{\left(\frac{c^x - 1}{c - 1}\right).b + c^x.v^2\right\}^{\frac{1}{2}}$$

Again, let $b = 0$, $c = 1$, and we easily find

$$f(v) = a + \sqrt{(v^2 - 2\,av)}$$

$$f^x(v) = a + \sqrt{\left(v^2 - 2\,av - \frac{0}{0}\right)} = a + \sqrt{\{v^2 - 2\,av - (x-1).a^2\}}$$

In this last expression, if we make $v = a = 1$, and for x put $-x$,

$$f(v) = 1 + \sqrt{(v^2 - 2v)}, \qquad f^{-x}(1) = 1 + \sqrt{(x)}.$$

We shall terminate this part of our subject with the following theorem: " *That an equation of differences, of whatsoever order and degree, (n and m,) with constant coefficients, and homogeneous in u_x, u_{x+1}, &c. is either completely integrable, or admits of one or more particular integrals, which may be found, their number not exceeding n.m.*" It is sufficient to substitute in it $C\,a^x$ for u_x, and we obtain an algebraic equation for determining a, whose degree will not exceed $n.m$. To fix our ideas, let us take the equation

$$0 = u^2_{x+2} + au_{x+2}.u_{x+1} + bu_{x+2}.u_x + c\,u^2_{x+1} + du_{x+1}.u_x + eu^2_x; \ldots \ldots \ldots \{40\}$$

The substitution of $C\,a^x$, or a^{x+C} for x, gives

$$0 = a^4 + a.a^3 + (b+c).a^2 + d.a + e; \ldots \ldots \ldots \ldots (f)$$

Let ^1a, ^2a,... be the roots of this equation, then will four particular integrals of $\{40\}$ be

$$u_x = {}^1C.{}^1a^x, \qquad u_x = {}^2C.{}^2a^x, \qquad u_x = {}^3C.{}^3a^x, \qquad u_x = {}^4C.{}^4a_x$$

and it is curious to observe, that the coefficients of $u_{x+2}u_x$, and u^2_{x+1}, are similarly involved in ^1a, ^2a, ^3a, ^4a. Let us now conceive the equation $\{40\}$ so constituted, that

$$0 = (u_{x+2} + A\,u_{x+1} + B\,u_x)(u_{x+2} + C\,u_{x+1} + D\,u_x)$$

2 A

which being evolved and compared with {40}, gives

$$A+C=a \left.\begin{array}{r}\end{array}\right\}, \qquad B+D=b \left.\begin{array}{r}\end{array}\right\}, \quad \text{and } BC+AD=d$$
$$AC=c \qquad\qquad BD=e$$

The four first of these determine A, B, C, D, and there remains an equation of condition

$$0 = d^2 + c b^2 + e a^2 - 4 c e - a b d.$$

If this be fulfilled, the resolution indicated is practicable, and {40} admits of two distinct complete integrals,

$$u_x = {}^1C \cdot {}^1a^r + {}^2C \cdot {}^2a^r$$

$$u_x = {}^3C \cdot {}^3a^x + {}^4C \cdot {}^4a^x$$

The equation of the first order and m^{th} degree

$$0 = u_{x+1}^m + {}^1A \cdot u_{x+1}^{m-1} \cdot u_x + \ldots \cdot {}^mA \cdot u_x^m$$

is always decomposable into factors of the first order and degree, and of course admits of m distinct complete integrals

$${}^1C \cdot {}^1a^x, \quad {}^2C \cdot {}^2a^x, \quad \&c.$$

There is one exception to our theorem, viz. when the sum of the products of the indices below each letter with the exponents above, respectively corresponding, in each several term, is the same throughout the equation. For instance, in the case of

$$0 = u_x u_{x+2} - a \cdot u_{x+1}^2, \qquad \text{where} \quad 1 \cdot (x+2) + 1 \cdot x = 2 \cdot (x+1)$$

the preceding method cannot be applied, the reason of which is plain.

There is again an exception to this exception, and that is, when the sum of all the coefficients of the equation $=0$; as in

$$0 = u_{x+1}^4 - 2 u_{x+2} \cdot u_{x+1}^2 \cdot u_x + u_{x+4} \cdot u_x^3$$

$$0 = u_{x+1}^n \cdot u_x^{m-n} + A \cdot u_{x+2} \cdot u_{x+1}^{n-2} \cdot u_x^{m-n+1} + B \cdot u_{x+2}^2 \cdot u_{x+1}^{n-4} \cdot u_x^{m-n+2} + C \cdot \&c. : \ldots \ldots \{41\}$$

where $\quad 0 = 1 + A + B + C + \&c.$

In such cases, the same substitution succeeds, but with this difference in the results, that while in the former the resulting equation determined the values of a; in these it is arbitrary, the whole vanishing by reason of the equation of condition.

In $\{41\}$, $u_x = a^{x+c}$, a and C being two arbitrary constants. This expression is consequently one of the complete integrals of $\{41\}$. To obtain the rest, we must consider that the elimination of a and C gives

$$0 = u_{x+1}^2 - u_{x+2} \cdot u_x$$

This must therefore hold good at the same time with $\{41\}$, and must consequently be a divisor of that equation. The quotient, made equal to nothing, will give the other integrals.

PART III.

ON FUNCTIONAL EQUATIONS.

$\mathbf{T}_{\text{HE}}$ integration of equations of partial differences, having introduced arbitrary functions of the independent variables, before such general integrals could be applied to any particular case, it was necessary to determine these functions, so as to satisfy, not only the general equation, but also the particular conditions of the problem, not expressed by that equation. Hence, arose a calculus, whose object might be stated to be, " the determination of functions from given conditions." Monge, in a Memoir published in the 5th Vol. of the **Melanges de Turin** (1770—1773), seems to have been the first who has treated this calculus by any thing like a regular process. He there resolves the equation

$$0 = \phi\{F(x)\} + A(x)$$

($F(x)$ and $A(x)$ being known functions of x, and ϕ being the characteristic of a function, whose form is required) and gives several instances of elimination, between two or more functional equations, which conduct to a final equation of this form. Lagrange, indeed, had previously * resolved the functional equation

$$a \cdot \phi\{t + a(h+kt)\} + \beta \cdot \phi\{t + b(h+kt)\} + \&c. = T$$

(T being any function of t, and a, β, ... &c. being constant;)—or, at least reduced it to the resolution of an exponential equation

$$0 = a \cdot (1+ka)^x + \beta(1+kb)^r + \&c.$$

but his process, (founded on the expansion of $\phi\{t + a(h+kt)\}$, &c. by Taylor's theorem, and the sub-

---

* In the third Vol. of the Melanges de Turin (1762—1765) in a Memoir entitled " Solutions de differens problemes de Calcul Integral.

sequent integration of the resulting differential equation, of an infinite order) is to be considered merely, as one of those casual and insulated applications of principles foreign to the subject, with which analysis abounds in every part. It was soon perceived, however, that all the cases of this calculus which could occur in the problems which gave rise to it, might be reduced to the integration of equations of finite differences, in which the difference of the independent variable, instead of being, as usual, unity, was some known function of the variable itself. This circumstance was first pointed out by Monge, who, following up his former ideas, published two Memoirs on the same subject, among those of the Savans Etrangers for 1773, in the latter of which, he solves several of the more simple cases, where the difference of the variable in the resulting equation is *proportional* to the variable itself, by ingenious, but peculiar artifices. But this was merely shifting the difficulty to another point. Such equations of differences had not been considered, and it became necessary to seek a general method of transforming them into equations of the common form, or of integrating them without such transformation. Precisely at this period, Laplace (not improbably at some suggestion of Monge) undertook this subject; and, if we may judge from an expression used by the latter *, considered the difficulty as completely overcome. Since that time, I am not acquainted with any thing of consequence that has been added to the discoveries of the above Authors. Nevertheless, Laplace's method, as he has delivered it, extends simply to the case he has considered, and no farther, viz. when the proposed functional equation involves two terms, $\phi\{^{(1)}F(x)\}$, and $\phi\{^{(2)}F(x)\}$, and two only.

I propose, in the following pages, to exhibit a method by which we may overcome the difficulty, not only for any number whatever of terms of the form

$$\phi\{^{(1)}F(x)\}, \quad \phi\{^{(2)}F(x)\}, \ldots\ldots\ldots \phi\{^{(n+1)}F(x)\},$$

in which case, the functional equation to be resolved, is

$$0 = F\left\{x, \phi\{^{(1)}F(x)\}, \quad \phi\{^{(2)}F(x)\}, \ldots \phi\{^{(n+1)}F(x)\}\right\}; \ldots\ldots\ldots \{42\}$$

but also, when the function to be found, is considered as relative to any number whatever of variables, $x, y, z, \ldots$ &c.; as for example, when it is required to determine the form of a function $\phi\{x, y, \ldots\}$, which shall satisfy the equation

$$0 = F\left\{x, y, \ldots, \phi\{^{(1)}F(x), {}^{(1)}f(y), \ldots\}, \ldots \phi\{^{(n+1)}F(x), {}^{(n+1)}f(y)\ldots\}\right\}; \ldots \{43\}$$

---

* ———" cet habile Géomêtre m'a dit qu'il convertissoit *toujours* une equation aux différences finies, et variables, en une équation aux différences finies et constantes."

F, $^{(1)}F, \ldots {}^{(n+1)}F$, $^{(1)}f, \ldots {}^{(n+1)}f$, &c. being the characteristics of given functions; a case, if I mistake not, which has never before been considered. It would be, however, a want of candour to omit acknowledging myself indebted to the above-mentioned method of Laplace for the first idea of the principle employed. I shall not scruple then to insert this method, (which is very short,) not only as an excellent introduction to what follows, but also as it will afford us subject for some remarks of considerable moment.

To begin, however, with the most simple case, let us take the equation

$$0 = \phi \{F(x)\} + A_x ; \ldots \ldots \ldots \ldots \{44\}$$

resolved by Monge. For x write * $F^{-1}(x)$, and we find

$$0 = \phi(x) + (AF^{-1})_x \quad \text{or,} \quad \phi(x) = -(AF^{-1}),$$

$$\text{since} \quad F\{F^{-1}(x)\} = x, \quad \text{or,} \quad FF^{-1} = 1.$$

Thus the nature of the function ϕ is determined; for F being a given characteristic, F^{-1} is given also by the resolution of the equation

$$F\{F^{-1}(x)\} = x$$

Let us next take the equation

$$0 = F\{x, \phi\{^{(1)}F(x)\}, \phi\{^{(2)}F(x)\}\} ; \ldots \ldots \ldots \ldots \{45\}$$

which is an extension of Laplace's case. To reduce it to an equation of common differences, assume

$$^{(1)}F(x) = u_z, \quad ^{(2)}F(x) = u_{z+1}$$

---

* Monge has, upon a different occasion, used $`\Phi(x)$ to denote the *inverse function* of $\Phi(x)$. Knight, in a paper on the expansion of any functions of Multinomials, (Phil. Trans. 1811. Part I.) uses " strokes put *under* a quantity to denote the reverse operation of strokes put *over* it." It is time, that such arbitrary notations should be replaced by one founded on some regular principle, which we have accordingly attempted to do. Analysts will judge whether our attempt has been attended with success. This process is in fact a demonstration of the following theorem. " If ϕ, A, B, represent any functional characteristics, such, that $\phi A(x)$, or $\phi\{A(x)\}$, $= B(x)$, or detaching the symbols of operation from those of quantity, $\phi A = B$, then we shall have, $\phi = BA^{-1}$, or $\phi x = B\{A^{-1}(x)\}$" In like manner, if $\phi AB = C$, we shall have $\phi = CB^{-1}A^{-1}$, for $\phi AB = (\phi A)B$, and consequently, considering (ϕA) as one characteristic by the preceding theorem, since $(\phi A)B = C$, $\phi A = CB^{-1} = (CB^{-1})$; and again, $\phi = (CB^{-1})A^{-1} = CB^{-1}A^{-1}$. In general, if $\phi . ^{(1)}A . ^{(2)}A \ldots ^{(n)}A = B$, we have $\phi = B . ^{(n)}A^{-1} . ^{(n-1)}A^{-1} \ldots ^{(1)}A^{-1}$, a theorem of great use, and which sets in a clear light the analogy between functional and exponential indices.

z being a certain function of x. The elimination of x from these, gives

$$0 = u_{z+1} - {}^{(2)}\mathrm{F} \, . \, {}^{(1)}\mathrm{F}^{-1}(u_z) \; ; \dots \dots \dots \dots \{46\}$$

an equation which *determines the characteristic u, in z*, or the form of the function u_z. It is an equation of common finite differences, (and we may observe, with none but constant coefficients) and, the theory of these equations being supposed perfect, u may be considered as a known characteristic. The equation $\{45\}$ becomes then

$$0 = F \{ ({}^{(1)}\mathrm{F}^{-1}u)_z, \quad (\phi u)_z, \quad (\phi u)_{z+1} \} \; ; \dots \dots \dots \dots \{47\}$$

an equation of common differences in z, the unknown characteristic being (ϕu). Let the value of $(\phi u)_z$ derived from this be w_z, and we obtain

$$\phi u = w, \quad \text{and} \quad \phi = w u^{-1}$$

Hence this theorem :——" To satisfy the functional equation

$$0 = F \{ x, \phi \{ {}^{(1)}\mathrm{F}(x) \}, \quad \phi \{ {}^{(2)}\mathrm{F}(x) \} \} \; ;$$

First, take u, a characteristic *determined in z* by the equation of common differences

$$0 = u_{z+1} - {}^{(2)}\mathrm{F} \, . \, {}^{(1)}\mathrm{F}^{-1}(u_z) :$$

Secondly, take w, a characteristic determined also in z, by the equation

$$0 = F \{ ({}^{(1)}\mathrm{F}^{-1}u_z), w_z, w_{z+1} \}$$

Then will the required characteristic ϕ, be expressed by the following combination of w, and u,

$$\phi = w u^{-1}$$

Now, on this process we have to remark, that in the integration of $\{46\}$ and $\{47\}$, there will be introduced two arbitrary functions of $\cos 2\pi z$; thus we shall have

$$u_z = \text{funct.} \{ z, \mathrm{C} (\cos 2\pi z) \} \; ; \dots \dots \dots (f)$$

C being an arbitrary characteristic.

Now the expression above found for ϕ, requires the determination of u^{-1}. For z then in the equation (f), write $u^{-1}(z)$, and we have

$$\text{funct.} \{ u^{-1}(z), \mathrm{C} (\cos 2\pi u^{-1}(z)) \} = u u^{-1}(z) = z$$

So long then as we particularize the characteristic C, the determination of $u^{-1}(x)$ in z, from this equation, depends only on the ordinary analysis, but when it is taken with the utmost regard to generality, so as to embrace as well discontinuous as analytical functions, we fall on a difficulty, which seems first to have been felt in the integration of partial differential equations *, and which is evidently insurmountable in any state of analysis. I mean, the general resolution of algebraic and transcendental equations, involving arbitrary functions of the unknown symbol. Laplace, in his Memoir, has taken no notice of this circumstance. He has indeed introduced an arbitrary function of $\cos 2\pi z$ in the integration of his equation, which corresponds to our $\{47\}$, but simply a constant in the other, where this difficulty takes its origin. In consequence of this omission, all the examples he has given of his method, although by no means limited, are yet very far short of that extreme generality, which the nature of the case admits.—We would not however be misunderstood. It would ill become us at any time, but more especially under the present circumstances, to criticise unguardedly on the models of admitted excellence. All that we intend by the preceding observation is, to add one more to the innumerable instances, where the simplicity of a remark has been the cause of its eluding notice, in the career of elaborate and successful research.

We come now to the general problem ; to determine the form of a function, whose characteristic is ϕ, so as to satisfy the condition

$$0 = F\left\{x,\ \phi\{^{(1)}F(x)\},\ \phi\{^{(2)}F(x)\},\ \ldots \phi\{^{(n+1)}F(x)\}\right\} ; \ldots\ldots\ldots\{42\}$$

$F,\ ^{(1)}F,\ \ldots\ ^{(n+1)}F$, being given characteristics. It is evident, that the foregoing method, in its present state, will not apply here ; for the equations

$$^{(1)}F(x) = u_z,\qquad ^{(2)}F(x) = u_{z+1},\ \ldots\ ^{(n+1)}F(x) = u_{z+n}$$

will not necessarily hold good at once †. But we shall be more successful, if we

---

* Euler. Inst. Calc. Int. tom. III. Probl. 12.

† If, however, this *should* happen, we need go no farther. For instance, in the equation noticed by Paoli, and delivered by Lacroix, in the third Vol. of his Diff. and Int. Calc. (Art. 989.)

$$0 = \underset{a^n x}{u} + {}^1A_x \cdot \underset{a^{n-1} x}{u} + \ldots\ldots{}^n A_x \cdot u_x + B_x$$

integrated by the former, in the case of $B_x = 0$, and ^1A_x &c. constant, by the substitution of $a\,x^\mu$ for u_x, and on which Lacroix remarks, rather too generally, " La même méthode," (celle de Laplace,) " s'applique sans difficulté aux equations des ordres superieurs."

proceed as follows. In the place of one variable z, assume n quantities $z_{(1)}$, $z_{(2)}, \ldots z_{(n)}$ functions of x, to be determined so as to satisfy the following system of equations

$$
\left.
\begin{aligned}
(1) \ldots \ldots \ldots {}^{(1)}\mathrm{F}(x) &= u_{z_{(1)}, \ z_{(2)}, \ldots \ldots z_{(n)}} \\
(2) \ldots \ldots \ldots {}^{(2)}\mathrm{F}(x) &= u_{z_{(1)}-1, \ z_{(2)}, \ldots \ldots z_{(n)}} \\
(3) \ldots \ldots \ldots {}^{(3)}\mathrm{F}(x) &= u_{z_{(1)}, \ z_{(2)}-1, \ldots \ldots z_{(n)}} \\
\cdot \quad \cdot \quad \cdot \quad \cdot \quad \cdot \quad \cdot \quad \cdot \quad \cdot \quad \cdot \quad \cdot \quad \cdot \quad \cdot \quad \cdot \\
(n+1) \ldots \ldots {}^{(n+1)}\mathrm{F}(x) &= u_{z_{(1)}, \ z_{(2)}, \ldots \ldots z_{(n)}-1}
\end{aligned}
\right\} ; \ldots \ldots \{48\}
$$

from which, eliminating x, we find

$$
\left.
\begin{aligned}
(1) \ldots \ldots \ldots {}^{(1)}\mathrm{F}^{-1} \left\{ u_{z_{(1)}, \ldots \ldots z_{(n)}} \right\} &= {}^{(2)}\mathrm{F}^{-1} \left\{ u_{z_{(1)}-1, \ z_{(2)}, \ldots \ldots z_{(n)}} \right\} \\
(2) \ldots \ldots \ldots {}^{(1)}\mathrm{F}^{-1} \left\{ u_{z_{(1)}, \ldots \ldots z_{(n)}} \right\} &= {}^{(3)}\mathrm{F}^{-1} \left\{ u_{z_{(1)}, \ z_{(2)}-1, \ldots \ldots z_{(n)}} \right\} \\
\cdot \quad \cdot \quad \cdot \quad \cdot \quad \cdot \quad \cdot \quad \cdot \quad \cdot \quad \cdot \quad \cdot \quad \cdot \quad \cdot \\
(n) \ldots \ldots \ldots {}^{(1)}\mathrm{F}^{-1} \left\{ u_{z_{(1)}, \ldots \ldots z_{(n)}} \right\} &= {}^{(n+1)}\mathrm{F}^{-1} \left\{ u_{z_{(1)}, \ z_{(2)}, \ldots \ldots z_{(n)}-1} \right\}
\end{aligned}
\right\} ; \ldots \ldots \{49\}
$$

These are equations of partial differences of the first order, whose integrals will be of the form

$$
\left.
\begin{aligned}
(1) \ldots u_{z_{(1)}, \ldots \ldots z_{(n)}} &= {}^{(1)}\mathrm{U} \left\{ z_{(1)}, \ldots z_{(n)}, {}^{(1)}\mathrm{C}(\cos 2\pi z_{(1)}, z_{(2)}, \ldots z_{(n)}) \right\} \\
(2) \ldots u_{z_{(1)}, \ldots \ldots z_{(n)}} &= {}^{(2)}\mathrm{U} \left\{ z_{(1)}, \ldots z_{(n)}, {}^{(2)}\mathrm{C}(z_{(1)}, \cos 2\pi z_{(2)}, \ldots z_{(n)}) \right\} \\
\cdot \quad \cdot \quad \cdot \quad \cdot \quad \cdot \quad \cdot \quad \cdot \quad \cdot \quad \cdot \quad \cdot \quad \cdot \quad \cdot \\
(n) \ldots u_{z_{(1)}, \ldots \ldots z_{(n)}} &= {}^{(n)}\mathrm{U} \left\{ z_{(1)}, \ldots z_{(n)}, {}^{(n)}\mathrm{C}(z_{(1)}, z_{(2)}, \ldots \cos 2\pi z_{(n)}) \right\}
\end{aligned}
\right\} ; \ldots \{50\}
$$

where ${}^{(1)}\mathrm{U}, \ldots {}^{(n)}\mathrm{U}$, are known, and ${}^{(1)}\mathrm{C}, \ldots {}^{(n)}\mathrm{C}$ arbitrary characteristics. Now we have

$$
x = \left({}^{(1)}\mathrm{F}^{-1} u \right)_{z_{(1)}, \ldots \ldots z_{(n)}}
$$

which substituted for x in $\{42\}$, gives

$$0 = F\left\{ \left(^{(1)}F^{-1}u\right)_{z_{(1)},\dots z_{(n)}},\ (\phi u)_{z_{(1)},\dots z_{(n)}},\ (\phi u)_{z_{(1)}-1,\dots z_{(n)}},\dots (\phi u)_{z_{(1)},\dots z_{(n)}-1} \right\};\dots\{51\}$$

an equation of partial differences of the usual form, in which (ϕu) is the unknown characteristic, and which is of the first order with respect to each of the independent variables. The integration of this will give us

$$(\phi u)_{z_{(1)},\dots\dots z_{(n)}} = \phi \left(u_{z_{(1)},\dots\dots z_{(n)}} \right) = \phi \left\{ ^{(1)}F(x) \right\} = w \left\{ z_{(1)},\dots z_{(n)},\ ^{1}\theta,\ ^{2}\theta,\ \&c. \right\}$$

w being the characteristic of a known function, and $^{1}\theta$, $^{2}\theta$, &c. being the arbitrary functions of the variables and of $\cos 2\pi z_{(1)}$, &c. introduced by the integration.

Now, by elimination from the n equations $\{50\}$, $z_{(1)},\dots z_{(n)}$ are given, (or may be conceived to be given) in functions of $u_{z_{(1)},\dots\dots z_{(n)}}$ or of $^{(1)}F(x)$. Conceive these values of $z_{(1)}\dots$ &c. to be substituted in the expression

$$w \left\{ z_{(1)},\dots z_{(n)},\ ^{1}\theta,\ ^{2}\theta,\ \&c. \right\}$$

and it will become a function of $^{(1)}F(x)$, which we will denote by $\bar{w}\, ^{(1)}F(x)$. Thus we have

$$\phi\, ^{(1)}F(x) = \bar{w}\, ^{(1)}F(x);\quad \text{and}\quad \phi = \bar{w}$$

This reduction of the equation $\{42\}$ to an equation of common differences, leaves nothing to desire in point of generality, at the same time that we must confess the operations required to be for the greater number of cases impracticable in the present state of analysis. This however, is no fault of the method itself, which, as we have said, supposes the absolute perfection of the ordinary analysis, and of the theory of finite differences. We may form some idea of the great generality of the resulting function, from the consideration that, independently of the arbitrary functions introduced in integrating $\{51\}$, the values of $z_{(1)},\ \dots z_{(n)}$ to be substituted in $w \{ z_{(1)},\dots z_{(n)},\ \&c.\}$ result from a series of eliminations between n equations, each of which contains an arbitrary function of $n-1$ of the quantities to be eliminated, and of the cosine of $2\pi x$ into the remaining one. Nor can any reason be assigned why these functions should be limited to analytical expressions, and not rather depend on no analytical law whatever, *or on a variable one* *.

We will now proceed to examine the general functional equation, involving two variables

$$0 = F\left\{x, y, \phi\left\{^{(1)}F(x), {}^{(1)}f(y)\right\}, \phi\left\{^{(2)}F(x), {}^{(2)}f(y)\right\}, \ldots \ldots \phi\left\{^{(n+1)}F(x), {}^{(n+1)}f(y)\right\}\right\}; \ldots \ldots \{43\}$$

F, $^{(1)}F$, $\ldots$ $^{(n+1)}F$, $^{(1)}f$ $\ldots$ $^{(n+1)}f$, being given characteristics.

Assume two series of quantities

$$z_{(1)}^{(1)}, \; z_{(2)}^{(1)}, \ldots . z_{(n)}^{(1)}, \quad \text{and,} \; z_{(1)}^{(2)}, \; z_{(2)}^{(2)}, \ldots . z_{(n)}^{(2)}$$

functions respectively of x and y, which shall satisfy the equations

$$\left\{\begin{array}{l} ^{(1)}F(x) = X_{z_{(1)}^{(1)},\, z_{(2)}^{(1)},\, \ldots z_{(n)}^{(1)}} \\[2mm] ^{(2)}F(x) = X_{z_{(1)}^{(1)}-1,\, z_{(2)}^{(1)},\, \ldots z_{(n)}^{(1)}} \\[2mm] \cdot \;\; \cdot \;\; \cdot \;\; \cdot \;\; \cdot \;\; \cdot \;\; \cdot \\[2mm] ^{(n+1)}F(x) = X_{z_{(1)}^{(1)},\, z_{(2)}^{(1)},\, \ldots z_{(n)}^{(1)}-1} \end{array}\right\} \text{and} \left\{\begin{array}{l} ^{(1)}f(y) = Y_{z_{(1)}^{(2)},\, z_{(2)}^{(2)},\, \ldots z_{(n)}^{(2)}} \\[2mm] ^{(2)}f(y) = Y_{z_{(1)}^{(2)}-1,\, z_{(2)}^{(2)},\, \ldots z_{(n)}^{(2)}} \\[2mm] \cdot \;\; \cdot \;\; \cdot \;\; \cdot \;\; \cdot \;\; \cdot \;\; \cdot \\[2mm] ^{(n+1)}f(y) = Y_{z_{(1)}^{(2)},\, z_{(2)}^{(2)},\, \ldots z_{(n)}^{(2)}-1} \end{array}\right\}; \ldots \left\{\begin{array}{l} 52, 1 \\ 52, 2 \end{array}\right\}$$

analogous to the system of equations $\{48\}$. From these, eliminate (as before) x and y, and we obtain

$$\left. \begin{array}{l} ^{(1)}F^{-1}\left\{X_{z_{(1)}^{(1)},\, \ldots . z_{(n)}^{(1)}}\right\} = {}^{(2)}F^{-1}\left\{X_{z_{(1)}^{(1)}-1,\, z_{(2)}^{(1)},\, \ldots . z_{(n)}^{(1)}}\right\} \\[3mm] ^{(1)}F^{-1}\left\{X_{z_{(1)}^{(1)},\, \ldots . z_{(n)}^{(1)}}\right\} = {}^{(3)}F^{-1}\left\{X_{z_{(1)}^{(1)},\, z_{(2)}^{(1)}-1,\, \ldots . z_{(n)}^{(1)}}\right\} \\[3mm] \cdot \;\; \cdot \;\; \cdot \;\; \cdot \;\; \cdot \;\; \cdot \;\; \cdot \;\; \cdot \;\; \cdot \;\; \cdot \;\; \cdot \;\; \cdot \\[3mm] ^{(1)}F^{-1}\left\{X_{z_{(1)}^{(1)},\, \ldots . z_{(n)}^{(1)}}\right\} = {}^{(n+1)}F^{-1}\left\{X_{z_{(1)}^{(1)},\, z_{(2)}^{(1)},\, \ldots . z_{(n)}^{(1)}-1}\right\} \end{array} \right\}; \ldots \ldots \{53, 1\}$$

---

* A discontinuous function of one variable, may be represented to the sense by the ordinate of a curve traced at random on a plane; of two, by that of a curve surface, any how spread forth in space. For more than two, we must have recourse to other considerations than those afforded by the modes of extension. We will add one remark. If any discontinuous function $\phi(x)$, be represented by the ordinate of such a curve, as we have described, then will $D\phi(x)$ or, $\dfrac{d\phi(x)}{d(x)}$ be represented by the ordinate divided by the subtangent, $\int dx . \left\{1 + (D\phi(x))^2\right\}^{\frac{1}{2}}$ by the arc, $D^{-1}\phi(x)$ by the area, and so forth, just as when ϕ is the characteristic of an analytical function.

and, similarly,

$$\left.\begin{aligned}
{}^{(1)}f^{-1}\left\{Y_{z^{(2)}_{(1)},\ldots z^{(2)}_{(n)}}\right\} &= {}^{(2)}f^{-1}\left\{Y_{z^{(2)}_{(1)}-1,\ z^{(2)}_{(2)},\ldots z^{(2)}_{(n)}}\right\}\\[6pt]
{}^{(1)}f^{-1}\left\{Y_{z^{(2)}_{(1)},\ldots z^{(2)}_{(n)}}\right\} &= {}^{(3)}f^{-1}\left\{Y_{z^{(2)}_{(1)},\ z^{(2)}_{(2)}-1,\ldots z^{(2)}_{(n)}}\right\}\\[6pt]
\cdot\quad\cdot\quad\cdot\quad\cdot\quad&\cdot\quad\cdot\quad\cdot\quad\cdot\quad\cdot\quad\cdot\quad\cdot\quad\cdot\quad\cdot\\[6pt]
{}^{(1)}f^{-1}\left\{Y_{z^{(2)}_{(1)},\ldots z^{(2)}_{(n)}}\right\} &= {}^{(n+1)}f^{-1}\left\{Y_{z^{(2)}_{(1)},\ z^{(2)}_{(2)},\ldots z^{(2)}_{(n)}-1}\right\}
\end{aligned}\right\}\ ;\ldots\ldots\{53, 2\}$$

The integration of these equations (being of partial differences) gives $X_{z^{(1)}_{(1)},\ldots z^{(1)}_{(n)}}$, and $Y_{z^{(2)}_{(1)},\ldots z^{(2)}_{(n)}}$ in the following forms :

$$X_{z^{(1)}_{(1)},\ldots z^{(1)}_{(n)}} = {}^{(1)}F(x) = {}^{(1)}U^{(1)}\left\{z^{(1)}_{(1)},\ldots z^{(1)}_{(n)},\ {}^{(1)}C^{(1)}(\cos 2\pi z^{(1)}_{(1)},\ z^{(1)}_{(2)},\ldots z^{(1)}_{(n)})\right\}$$

$$X_{z^{(1)}_{(1)},\ldots z^{(1)}_{(n)}} = {}^{(1)}F(x) = {}^{(2)}U^{(1)}\left\{z^{(1)}_{(1)},\ldots z^{(1)}_{(n)},\ {}^{(2)}C^{(1)}(z^{(1)}_{(1)},\ \cos 2\pi z^{(1)}_{(2)},\ldots z^{(1)}_{(n)})\right\}$$

$$\cdot\quad\cdot\quad\cdot\quad\cdot\quad\cdot\quad\cdot\quad\cdot\quad\cdot\quad\cdot\quad\cdot\quad\cdot\quad\cdot\quad\cdot\quad\cdot$$

$$X_{z^{(1)}_{(1)},\ldots z^{(1)}_{(n)}} = {}^{(1)}F(x) = {}^{(n)}U^{(1)}\left\{z^{(1)}_{(1)},\ldots z^{(1)}_{(n)},\ {}^{(n)}C^{(1)}(z^{(1)}_{(1)},\ z^{(1)}_{(2)},\ldots\cos 2\pi z^{(1)}_{(n)})\right\}$$

$$Y_{z^{(2)}_{(1)},\ldots z^{(2)}_{(n)}} = {}^{(1)}f(y) = {}^{(1)}U^{(2)}\left\{z^{(2)}_{(1)},\ldots z^{(2)}_{(n)},\ {}^{(1)}C^{(2)}(\cos 2\pi z^{(2)}_{(1)},\ z^{(2)}_{(2)},\ldots z^{(2)}_{(n)})\right\}$$

$$Y_{z^{(2)}_{(1)},\ldots z^{(2)}_{(n)}} = {}^{(1)}f(y) = {}^{(2)}U^{(2)}\left\{z^{(2)}_{(1)},\ldots z^{(2)}_{(n)},\ {}^{(2)}C^{(2)}(z^{(2)}_{(1)},\ \cos 2\pi z^{(2)}_{(2)},\ldots z^{(2)}_{(n)})\right\}$$

$$\cdot\quad\cdot\quad\cdot\quad\cdot\quad\cdot\quad\cdot\quad\cdot\quad\cdot\quad\cdot\quad\cdot\quad\cdot\quad\cdot\quad\cdot\quad\cdot$$

$$Y_{z^{(2)}_{(1)},\ldots z^{(2)}_{(n)}} = {}^{(1)}f(y) = {}^{(n)}U^{(2)}\left\{z^{(2)}_{(1)},\ldots z^{(2)}_{(n)},\ {}^{(n)}C^{(2)}(z^{(2)}_{(1)},\ z^{(2)}_{(2)},\ldots\cos 2\pi z^{(2)}_{(n)})\right\}$$

${}^{(1)}U^{(1)},\ldots {}^{(n)}U^{(1)},\ {}^{(1)}U^{(2)},\ldots {}^{(n)}U^{(2)}$ being determinate and given, but ${}^{(1)}C^{(1)},\ldots {}^{(n)}C^{(1)}$, ${}^{(1)}C^{(2)},\ldots {}^{(n)}C^{(2)}$, arbitrary characteristics. From these equations, $z^{(1)}_{(1)},\ldots z^{(1)}_{(n)}$, and $z^{(2)}_{(1)},\ldots z^{(2)}_{(n)}$ are given by elimination in functions respectively of ${}^{(1)}F(x)$, and of ${}^{(1)}f(y)$. This operation which we shall have occasion to refer to, let us denote by $\{A\}$: let us also express

$$\phi\left\{X_{z^{(1)}_{(1)},\ldots z^{(1)}_{(n)}},\ Y_{z^{(2)}_{(1)},\ldots z^{(2)}_{(n)}}\right\}$$

by the functional symbol

$$w_{x^{(1)}_{(1)}, \ldots x^{(1)}_{(n)},\ x^{(2)}_{(1)}, \ldots x^{(2)}_{(n)}}$$

for it is evidently a certain function both of $x^{(1)}_{(1)}, \ldots x^{(1)}_{(n)}$, and of $x^{(2)}_{(1)}, \ldots x^{(2)}_{(n)}$. The proposed equation $\{43\}$, becomes then

$$= F\left\{ ({}^{(1)}F^{-1}X)_{x^{(1)}_{(1)} \ldots x^{(1)}_{(n)}},\ ({}^{(1)}f^{-1}Y)_{x^{(2)}_{(1)}, \ldots x^{(2)}_{(n)}},\ w,\ w_{x^{(1)}_{(1)}-1,\ x^{(2)}_{(1)}-1},\ w_{x^{(1)}_{(2)}-1,\ x^{(2)}_{(2)}-1}, \ldots \ldots w_{x^{(1)}_{(n)}-1,\ x^{(2)}_{(n)}-1} \right\}$$

where, in the indices of w, all the letters that have not varied, are omitted for the sake of brevity. This is again an equation of partial differences of the common form, involving $2n$ variables, but, as it happens that these vary throughout the equation by pairs, and similarly, it may be reduced to another with only n. Let us now call the value of $w_{x^{(1)}_{(1)}} \ldots$ &c. deduced from this

$$w\ \left\{ x^{(1)}_{(1)}, \ldots x^{(1)}_{(n)},\ x^{(2)}_{(1)}, \ldots x^{(2)}_{(n)},\ {}^{1}\theta,\ {}^{2}\theta,\ \&c. \right\}$$

${}^{1}\theta$, ${}^{2}\theta$, &c. being the arbitrary functions, &c. introduced by the integration.

Now for $x^{(1)}_{(1)}, \ldots$ &c., and $x^{(2)}_{(1)}, \ldots$ &c. substitute their values in functions of ${}^{(1)}F(x)$, and ${}^{(1)}f(y)$, found by the operation $\{A\}$, and this expression will become a function of ${}^{(1)}F(x)$, and ${}^{(1)}f(y)$, which we will denote by the characteristic $\bar{w}$. We have then

$$\phi\left\{ {}^{(1)}F(x),\ {}^{(1)}f(y) \right\} = w_{x^{(1)}_{(1)}, \ldots} = \bar{w}\left\{ {}^{(1)}F(x),\ {}^{(1)}f(y) \right\}$$

and consequently,

$$\phi = \bar{w}$$

We have said enough to indicate the method of proceeding, whatever be the number of symbols, x, y, z, &c., relative to which the unknown characteristic ϕ is taken.

It will not be uninteresting to examine how far this theory may be applied to cases, which may actually occur, so as to fall on none but integrable equations. Now, it is easily seen, that any functional equation of the form

$$= \phi\left\{ {}^{(1)}F(x),\ {}^{(1)}f(y), \ldots \right\} + {}^{1}A.\,\phi\left\{ {}^{(2)}F(x),\ {}^{(2)}f(y), \ldots \right\} + \ldots {}^{n}A.\,\phi\left\{ {}^{(n+1)}F(x),\ {}^{(n+1)}f(y), \ldots \right\} + B\ ; \ldots \{54\}$$

2 D

in which B, ^{1}A,....^{n}A, are constant, leads to an equation of partial differences, integrable by Laplace's process *. This functional equation may therefore be resolved, provided we can integrate the preliminary equations of the first order, which may always be treated as equations of simple differences with only one variable, and with constant coefficients. A vast variety of cases might be imagined in which this may be accomplished, among which we may enumerate the following:

First, when $^{(1)}$F,....&c., $^{(1)}f$,....&c., &c., are the characteristics of rational algebraic functions of the first degree; which comprehends the equations

$$0 = \phi\left\{\frac{^0a + ^0b\,x}{^0c + ^0d\,x}\right\} + {}^1A.\phi\left\{\frac{^1a + ^1b\,x}{^1c + ^1d\,x}\right\} + \ldots\ldots {}^nA.\phi\left\{\frac{^na + ^nb\,x}{^nc + ^nd\,x}\right\} + B\,;\ldots\ldots\{55,1\}$$

$$0 = \phi\left\{\frac{^0a + ^0b\,x}{^0c + ^0d\,x}, \frac{^0a + ^0\beta\,y}{^0\gamma + ^0\delta\,y}\right\} + \ldots\ldots {}^nA.\phi\left\{\frac{^na + ^nb\,x}{^nc + ^nd\,x}, \frac{^na + ^n\beta\,y}{^n\gamma + ^n\delta\,y}\right\} + B\,;\ldots\{55,2\}$$

&c. = &c.

for, in this case, the general form of the preliminary equations is

$$0 = u_{\substack{z+1\\(i)}} . u_{\substack{z\\(i)}} + u_{\substack{z+1\\(i)}} . \left\{\frac{^ia . ^0d - ^ib . ^0c}{^id . ^0c - ^ic . ^0d}\right\} + u_{\substack{z\\(i)}} . \left\{\frac{^ic . ^0b - ^id . ^0a}{^id . ^0c - ^ic . ^0d}\right\} + \left\{\frac{^ib . ^0a - ^ia . ^0b}{^id . ^0c - ^ic . ^0d}\right\};$$

which we have previously shewn how to integrate. The details of this case would be found highly interesting, were it not for the necessity of studying compression as far as possible, to allow room for a few words on objects of superior importance. There are two cases, however, which afford a remarkable simplification; the first is, when any, or all of the following equations hold good,

$$\frac{^0c}{^0d} = \frac{^1c}{^1d}, \quad \frac{^0c}{^0d} = \frac{^2c}{^2d}, \ldots.\&c.\,; \qquad \frac{^0a}{^0b} = \frac{^1a}{^1b}, \quad \frac{^0a}{^0b} = \frac{^2a}{^2b}, \quad \&c. \quad \&c$$

the second, when

$$^0c = {}^1c = {}^2c = \&c. = 1, \quad \text{and} \quad ^0d = {}^1d = \&c. = 0.$$

---

* Savans Etrangers. 1773. Probls. VI. and IX. pages 90. and 110. Cousin. Lecons de Calc. Diff. et de Calc. Int. Part II. page 735. et suiv.

and similarly, for ^0a, ^1a, &c., $^0\beta$, &c., &c. The functional equations then take the forms

$$0 = \phi\{^0a + ^0b\,x\} + {}^1A\cdot\phi\{^1a + ^1b\,x\} + \ldots\ldots {}^nA\cdot\phi\{^na + ^nb\,x\} + B\,;\ldots\ldots\{55,3\}$$

$$0 = \phi\{^0a + ^0b\,x,\ ^0a + ^0\beta\,y\} + \ldots\ldots {}^nA\cdot\phi\{^na + ^nb\,x,\ ^na + ^n\beta\,y\} + B\,;\ldots\ldots\{55,4\}$$

The second general case which admits of solution is, when $^{(1)}F$, &c., $^{(1)}f$, &c. are the characteristics of functions of the form $a\,x^b$, which embraces such equations as

$$0 = u_{x^1} + {}^1A\cdot u_{x^2} + {}^2A\cdot u_{x^3} + \ldots\ldots {}^nA\cdot u_{x^{n+1}} + B\,;\ldots\ldots\ldots\{56,1\}$$

$$0 = \phi\{^0a\,x^{^0b}\} + {}^1A\cdot\phi\{^1a\,x^{^1b}\} + \ldots\ldots {}^nA\cdot\phi\{^na\,x^{^nb}\} + B\,;\ldots\{56,2\}$$

$$0 = \phi\{^0a\,x^{^0b},\ ^0a\,y^{^0\beta}\} + \ldots\ldots {}^nA\cdot\phi\{^na\,x^{^nb},\ ^na\,y^{^n\beta}\} + B\,;\ldots\ldots\{56,3\}$$

&c. = &c.

The third case is, when $^{(1)}F(x)$, &c. are of the form $a\cdot c^x$, as in the equation

$$0 = \phi\{^0a\cdot ^0c^x\} + \ldots\ldots {}^nA\cdot\phi\{^na\cdot ^nc^x\} + B\,;\ldots\ldots\ldots\{57\}$$

We must here notice also, that any combination of the above forms, after the manner of the following instance, is susceptible of resolution ;

$$0 = \phi\left\{\frac{^0a + ^0b\,x}{^0c + ^0d\,x},\ ^0a\,y^{^0\beta},\ ^0\gamma\cdot ^0\delta^z\right\} + \ldots {}^nA\cdot\phi\left\{\frac{^na + ^nb\,x}{^nc + ^nd\,x},\ ^na\,y^{^n\beta},\ ^n\gamma\cdot ^n\delta^z\right\} + B\,;\ldots\{58\}$$

The reason of which is evident, on a mere inspection of the equations $\{53,1\}$, and $\{53,2\}$.

There is a case which we have not yet considered, viz. when the unknown function of a simple constant is involved in the equation, as in the instance

$$0 = \phi\{x^2\} - \phi(x) - \phi(a).$$

Now, whatever be the form of the function ϕ, it is evident, that $\phi(a)$ will be independent on x, and consequently, if we suppose (in the case before us)

$$\dot{u}_z = x,\quad u_{z+1} = x^2 = u_z^2$$

independent also on z.   Thus, in the equation

$$0 = \phi\left(u_{z+1}\right) - \phi\left(u_z\right) - \phi\left(a\right)$$

$\phi\left(a\right)$ must be considered as constant, which gives

$$\left(\phi u\right)_z = z \,.\, \phi\left(a\right) + f\left\{\cos 2\,\pi\,z\right\}$$

Thus,

$$\phi\left(x\right) = \phi\left(a\right) \,.\, u^{-1}\left(x\right) + f\left\{\cos 2\pi\,u^{-1}\left(x\right)\right\}\; ;\ldots\ldots\ldots\ldots(g)$$

Meanwhile, the equation $u_{z+1} = u_z^2$, gives

$$u_z = \left\{\mathrm{F}\left(\cos 2\,\pi\,z\right)\right\}^{2^z}$$

To avoid complication, we shall take only a particular case of this, *viz.*
when $\mathrm{F}\left\{\cos 2\,\pi\,z\right\} = $ constant $= c$, or, $u_z = c^{2^z}$, and writing for z, $u^{-1}\left(x\right)$,

$$u\,u^{-1}\left(x\right) = x = c^{2^{u^{-1}(x)}},\quad\text{and}\quad u^{-1}\left(x\right) = \frac{\log^2 x - \log^2 c}{\log 2}$$

which substituted in (g),   gives

$$\phi\left(x\right) = \phi\left(a\right) \,.\, \frac{\log^2 x - \log^2 c}{\log 2} + f\left\{\cos\frac{2\,\pi\left(\log^2 x - \log^2 c\right)}{\log 2}\right\}$$

Now, let $x = a,$ and we have

$$\phi\left(a\right) = \phi\left(a\right) \,.\, \frac{\log^2 a - \log^2 c}{\log 2} + f\left\{\cos 2\,\pi \,.\, \frac{\log^2 a - \log^2 c}{\log 2}\right\}$$

which gives

$$\phi\left(a\right) = \frac{\log 2}{\log 2 - \log^2 a + \log^2 c} \,.\, f\left\{\cos 2\,\pi \,.\, \frac{\log^2 a - \log^2 c}{\log 2}\right\}$$

from which, we obtain

$$\phi\left(x\right) = \frac{\log^2 x - \log^2 c}{\log 2 - \log^2 a + \log^2 c} \,.\, f\left\{\cos 2\,\pi \,.\, \frac{\log^2 a - \log^2 c}{\log 2}\right\} + f\left\{\cos 2\,\pi \,.\, \frac{\log^2 x - \log^2 c}{\log 2}\right\}$$

and the simplest form of $\phi\left(x\right)$, which answers the condition, is

$$\phi\left(x\right) = \frac{\log^2 x - \log^2 a + \log 2}{\log 2 - \log^2 a}$$

We have been thus explicit in stating the preceding operations, in order to shew clearly, that *we have no right to consider* $\phi(a)$ *in such an equation as the proposed, in the light of an independent arbitrary constant;* its value being determined by the nature of the function f, and by the equation which arises from the substitution of a for x, in the value of $\phi(x)$. The same considerations will apply to any functional equation with one variable, involving any number of quantities, such as $\phi(a)$, $\phi(b)\ldots$; or to a functional equation in more variables than one, involving expressions of the form

$$\phi(a,\ b,\ c,\ldots), \quad \phi(a',\ b',\ c',\ldots), \&c.$$

Should expressions, such as $\phi(a, y)$, $\phi(x, b, z,\ldots)$, &c. enter into the proposed equation, we must, in like manner, consider them as determinate, but unknown functions, and, according to the principles above delivered, deduce an expression for $\phi\{x, y, z,\ldots\}$, of the utmost possible generality, and which will involve these particular functions.

To fix our ideas, let us conceive only two variables, x, y, and one particular function, $\phi(a, y)$. We shall have then

$$\phi(x, y) = F\{x, y, \phi(a, y), {}^{1}D\phi(a, y), {}^{2}D\phi(a, y),\ldots\} ;\ldots\ldots(h)$$

${}^{1}D\phi(a, y)$, ${}^{2}D\phi(a, y)$, &c. denoting functions of a and y, *any how derived,* according to known laws, from $\phi(a, y)$. The supposition $x = a$, gives

$$\phi(a, y) = F\{a, y, \phi(a, y), {}^{1}D\phi(a, y), {}^{2}D\phi(a, y),\ldots\}$$

a functional equation in one variable, which suffices for the determination of $\phi(a, y)$. The value of this, so found, being substituted in (h), gives the complete expression of $\phi(x, y)$.

We shall now proceed to consider cases where not only the unknown function $\phi(x, y,\ldots)$, but, also its differential coefficients of any order, taken with respect to one or more of the variables, are involved. Let us conceive, (in the case of one variable x,) that the proposed equation contains such expressions as

$$D\phi\{F(x)\}, \quad D^{2}\phi\{F(x)\}, \&c.$$

2 E

After making substitutions similar to those already described, the equation will, besides the terms

$$(\phi u)_{z_{(1)}, \ldots z_{(n)}}, \ (\phi u)_{z_{(1)}-1, \ldots z_{(n)}}, \ \&c.$$

contain functional terms of the forms $(D\phi)u$, $(D^2\phi)u$, ... where in the derivations x is considered as the independent variable, and which it remains to transform into such expressions as $D(\phi u)$, $D^2(\phi u)$, &c. so as to reduce the equation to one of mixed differences, where (ϕu) is the characteristic of the unknown function. Now this is easily accomplished, as follows:

$$(D_x \phi) u = \frac{d(\phi u)}{d u} = \frac{1}{D_x u} D_x (\phi u)$$

$$(D_x^2 \phi) u = \frac{d\{(D_x \phi)u\}}{d u} = \frac{1}{D_x u} D_x \cdot \frac{1}{D_x u} D_x (\phi u)$$

$$(D_x^3 \phi) u = \frac{d\{(D_x^2 \phi)u\}}{d u} = \frac{1}{D_x u} D_x \cdot \frac{1}{D_x u} D_x \cdot \frac{1}{D_x u} D_x (\phi u)$$

$$\&c. \quad = \quad \&c. \quad = \quad \&c.$$

where it must be observed, that ψ being the characteristic of any function of x, or, of any implicit function of $z_{(1)}, \ldots z_{(n)}$, we have

$$D_x \psi = \frac{d \psi}{d x} = \frac{\left(\dfrac{d\psi}{d z_{(1)}}\right) d z_{(1)} + \left(\dfrac{d\psi}{d z_{(2)}}\right) d z_{(2)} + \&c.}{\left(\dfrac{d x}{d z_{(1)}}\right) d z_{(1)} + \left(\dfrac{d x}{d z_{(2)}}\right) d z_{(2)} + \&c.}$$

With regard to the denominator,

$$\left(\frac{d x}{d z_{(1)}}\right) d z_{(1)} + \left(\frac{d x}{d z_{(2)}}\right) d z_{(2)} + \&c.$$

no difficulty can arise, since x is given in functions of $z_{(1)}, \ldots$ by the preliminary equations, as in {48}, and u being, as in the former cases, the characteristic of a known function in these variables, we may thus readily express $\dfrac{1}{D_x u} D_x \phi(u)$, &c. in the required form. Similar considerations will apply, when the unknown function ϕ is taken with respect to more than one variable $x, y, z, \ldots$ &c. but perhaps it is needless, in the present state of analysis, to push our enquiries farther on this head.

Thus we may regard the theory of functional equations of the first order as complete, whatever be the number of variables they involve. By equations of the first order, I mean, such as involve simply the unknown function ϕ, and its derived functions any how combined with the variables $x, y, \ldots$ &c. Such equations as involve at the same time ϕ, and ϕ^2, or $\phi\phi$, and their derived functions, may with great propriety be termed functional equations of the second order, and so on. For example, the equations

$$\{\phi(x)\}^2 = \phi(2x) + 2$$

$$\phi(x^q) = \phi(mx) + p$$

solved by **Laplace**, as well as

$$\phi\left\{\frac{B u^m}{(b-A)^m}\right\} - \phi\left\{\frac{D u^n}{(b-C)^n}\right\} = \frac{(a-b)(A-C)}{(b-A)(b-C)} \cdot u$$

integrated by **Monge**, are of the first order. The following

$$0 = \phi^2(x) + A_x \cdot \phi(x) + B_x$$

$$0 = \{\phi(a+bx+cx^2)\}^m + A_x \cdot \phi^2\left\{\frac{d+ex}{f+gx}\right\} + B_x \cdot \frac{d^n \phi(x)}{dx^n} + C_x$$

$$0 = A_x + B_x \cdot \Sigma \phi^2(x) + C_x \cdot \int dx \cdot \phi(x-x^2)$$

are of the second. Such equations also as

$$\phi\{A_x + B_x \cdot \phi(C_x)\} = E_x$$

may be classed among equations of the second order, as, by taking the function ϕ^{-1} on both sides, we find

$$0 = A_x + B_x \cdot \phi(C_x) - \phi^{-1}(E_x)$$

where the difference between the greatest and least indices of ϕ, is 2.

The theory of functional equations of the second and superior orders, appears extremely difficult, and at present we can hazard nothing general on the subject, farther than simply this.   Conceive $\phi(x)$ to be a certain function of x, and any number of constants, a, b, c, ...&c.

From this expression, conceive $\phi^2(x)$, $\phi^3(x)$, ...$\phi^n(x)$, to be successively formed, which will of course be functions of x, a, b, c, &c. and we may therefore suppose n such equations as the following to exist cotemporaneously

$$
\left.
\begin{aligned}
(1) &\ldots\ldots \phi(x) = {}^1f\{x, a, b, \ldots,\} \\
(2) &\ldots\ldots \phi^2(x) = {}^2f\{x, a, b, \ldots\} \\
&\ldots\ldots\ldots\ldots\ldots\ldots\ldots\ldots\ldots\ldots\ldots \\
(n) &\ldots\ldots \phi^n(x) = {}^nf\{x, a, b, \ldots\}
\end{aligned}
\right\} ;\ldots\ldots\ldots\ldots (i)
$$

From these equations, it would be possible to eliminate $(n-1)$ of the quantities a, b, c, ..., and since the resulting equation

$$0 = \mathrm{F}\{x, \phi(x), \phi^2(x), \ldots \phi^n(x)\} ;\ldots\ldots\ldots\ldots \{59\}$$

is independent of the quantities so eliminated, and consequently the same, whatever values we assign to them, they are in fact arbitrary, so far as regards this equation.   In reascending then from the equation $\{59\}$ to the complete expression of $\phi(x)$, (which we may be allowed to call its integral,) we must introduce $(n-1)$ arbitrary constants.   Again, suppose the elimination to have taken place only between m of the equations (i); $\big((i, n)$ being one of them,$\big)$ and m any number less than n, and the resulting equation will contain m of the functional terms $\phi(x), \ldots \phi^n(x)$.   At the same time, $m-1$ of the constants a, b, c, ... will have disappeared.   Hence this theorem; that, *the integral of a functional equation*, of whatever order, *containing m functional terms, in which the quantities under the several characteristics ϕ, ϕ^2, ... are the same throughout, involves m − 1 arbitrary constants; and reciprocally, if an expression for the unknown function can be found, involving m − 1 irreducible arbitrary constants, it is the complete integral.*

Let us suppose, for example, $\phi(x) = a + bx$, an equation which gives

$$\phi^2(x) = a(1+b) + b^2 x$$

$$\phi^3(x) = a(1+b+b^2) + b^3 x.$$

Eliminating a from the two first, we find

$$0 = \phi^2(x) - (1+b) \cdot \phi(x) + bx.$$

Had we eliminated b, we should have had

$$0 = \phi^2(x) + \frac{a \cdot \phi(x) - \{\phi(x)\}^2}{x} - a.$$

Again, making use of all three equations to eliminate both a and b,

$$0 = \frac{(\phi^2 x - \phi x)^3}{\phi x - x} - (\phi x)^2 + x \cdot \phi^2 x + (2\phi x - \phi^2 x - x) \cdot \phi^3 x$$

If we suppose $\phi(x, y, z, \ldots)$ to denote any function whatever of $x, y, z, \ldots$, and conceive this expression substituted for x, thus,

$$\phi\{\phi(x, y, z, \ldots), y, z, \ldots\}$$

we shall have the second *partial function*, taken with respect to x only; if this be again substituted for x, the third, and so on. In like manner, we may form the successive partial functions with respect to $y, z, \ldots$. If the mth partial function with respect to x, be in the same manner continually substituted n times for y, in the expression

$$\phi(x, y, z, \ldots)$$

we shall obtain a result, which we will denote by

$$\phi^{m, n, 1, \ldots}(x, y, z, \ldots)$$

and so on to

$$\phi^{m, n, p, \ldots}(x, y, z, \ldots)$$

This notation is indeed imperfect, as it will not express any variations we may make in the *order* of taking the functions, which is not here, as in the theory of partial differentials, a matter of indifference: but it will suffice for our immediate purpose which is merely to indicate the existence of a calculus of partial functions of a description entirely new.

2 F

The following are equations of partial functions:

$$0 = \phi^{2,1}\{x, y\} + A_{x,y} \cdot \phi^{1,2}\{x, y\} + B_{x,y}$$

$$0 = \phi^{2,1,1}\{x, y, z\} + A_{x,y,z} \cdot \phi^{1,2,1}\{x, y, z\} + \&c.$$

A question may arise, from what we have already seen of the manner in which arbitrary functions enter into the integrals of functional equations ; whether the application of such equations be really sufficient, entirely to determine the forms of arbitrary functions contained in the integrals of equations of partial differentials, by particular assigned conditions. Nor, indeed, does it seem possible to demonstrate that, from such integrals, after going through the operations above described, (or rather conceiving them to have been gone through,) with the utmost regard to generality, every arbitrary term should necessarily, and in all cases, vanish of itself. Should they not, new conditions must be assigned, and new functional equations resolved, whose integration will introduce fresh arbitrary functions, and so on, without the prospect of an end. The *denouement* of these difficulties seems reserved for a far more advanced state of Analytical Science, than we can at present boast of having attained.